The AMS Weather Book

The AMS Weather Book:
The Ultimate Guide to America's Weather

Jack Williams

Copublished with the American Meteorological Society

The University of Chicago Press
Chicago and London

Jack Williams was weather editor of *USA TODAY* for twenty-three years before becoming public
outreach coordinator for the American Meteorological Society in 2005. He has reported on weather
and climate research from Antarctica, Greenland, a research icebreaker on the Arctic Ocean, flights into
hurricanes, and tornado chases with scientists. He is the author of *The USA TODAY Weather Book, The USA
TODAY Weather Almanac, The Complete Idiot's Guide to the Arctic and Antarctic*, and co-author with
Dr. Bob Sheets of *Hurricane Watch: Forecasting the Deadliest Storms on Earth*.
He and his wife, Darlene, live in Falls Church, Virginia.

The American Meteorological Society seeks to advance the atmospheric and related
sciences, technologies, applications, and services for the benefit of society.

The University of Chicago Press, Chicago 60637
The University of Chicago Press, Ltd., London
© 2009 by the American Meteorological Society
All rights reserved. Published 2009
Printed in the United States of America

18 17 16 15 14 13 12 11 10 09   1 2 3 4 5

ISBN-13: 978-0-226-89898-8 (cloth)

ISBN-10: 0-226-89898-9 (cloth)

Front endpaper: *Lightning*, © University Corporation for Atmospheric Research.
Back endpaper: *Rainbow*, © University Corporation for Atmospheric Research, photo by Bob Henson.

Library of Congress Cataloging-in-Publication Data

Williams, Jack.
The AMS weather book : the ultimate guide to America's weather / Jack Williams.
p. cm.
"Copublished with the American Meteorological Society."
ISBN-13: 978-0-226-89898-8 (cloth : alk. paper)
ISBN-10: 0-226-89898-9 (cloth : alk. paper) 1. Weather—Popular works.
2. Climatology—Popular works. 3. Meteorology—Popular works. I. Title.
QC981.2.W55 2009
551.5—dc22
2008035916

♾The paper used in this publication meets the minimum
requirements of the American National Standard for Information Sciences—Permanence of Paper
for Printed Library Materials, ANSI Z39.48-1992.

CONTENTS

A brief look at the topics

and themes you'll find in the

coming pages

"Everybody talks about the weather, but nobody does anything about it." How many times have you heard that saying? The first part of it is true. Everyone talks about the weather because it affects all of us all the time. But as you will read in the chapters ahead, there are many people who are doing a lot about the weather by conducting research, developing increasingly accurate computer forecasting models, and creating better and more sensitive instruments to monitor the atmosphere. Storms like Hurricane Katrina, as a recent and dramatic example, represent the violent side of the atmosphere and strike fear into our hearts, but the only remedy for fear is knowledge. This is the book for those who want to learn more about how the weather works and enjoy the process.

After Hurricane Katrina hit on August 29, 2005, the photograph on the previous two pages and others like it began popping up on Web sites and landing in e-mail boxes. The caption identified it as Hurricane Katrina making landfall. A few weeks later, while I was eating lunch in the National Center for Atmospheric Research (NCAR) cafeteria in Boulder, Colorado, someone mentioned these photos. We had a good laugh about how people will believe anything. The photo is obviously a powerful Great Plains thunderstorm, not a hurricane.

I realized later, however, that the misidentification is not really all that obvious unless you know more about weather than most people.

That realization helped crystallize one of my thoughts about this book, then just the glimmer of an idea I had discussed with the American Meteorological Society (AMS), and what it should accomplish. That is, it should help readers look at the sky (or photographs of the sky) with an informed appreciation of what they're seeing. It should also help readers understand news about weather and climate, cope with weather threats (including the different dangers presented by hurricanes and the storm in the photo), and learn how the atmospheric and oceanographic sciences are a part of the story of human understanding of the physical world.

This book is written for anyone who is curious and excited about weather and how the atmosphere works. Instead of writing a textbook that covers only hard science, we developed a book that focuses on the human side of the atmospheric sciences. It includes stories about people coping with weather events or working to improve understanding or forecasts of them, as well as a number of brief profiles of men and women whose professional lives focus on weather, oceanography, or climate. We also have 123 explanatory graphics, because the old saying about a picture being worth a thousand words applies to many scientific concepts.

The book uses a historical approach to explore topics, ranging from the blue sky in Chapter 1 to the ozone hole in Chapter 12, because the story of how scientists came to understand a phenomenon can help you understand it.

With a subject as big as Earth's atmosphere and oceans, we cannot go into immense detail on the many related topics addressed. To help you further these explorations, this book's accompanying Web site at http://www.amsweatherbook.com/ has links to other sites and recommendations for books and articles that further examine the topics in this book, as well as related topics that the book doesn't cover. Information usually found in footnotes is also on the Web site.

Now back to the photo in question. Two of my associates weighed in that day in Boulder: Peggy LeMone, an NCAR scientist profiled in Chapter 2, and Bob Henson, who's with NCAR media relations and author of the *Rough Guide to Weather 2* and the *Rough Guide to Climate Change 2*. Both agreed that in a hurricane, visibility wouldn't be as good as in this photo, because hurricanes are very humid, with haze, low clouds, and rain restricting visibility. Also, clouds in hurricanes don't have the structure of the cloud in the photo. The cloud that's shown almost touching the ground is a **wall cloud**. (Note that boldface type indicates the book's first use of a word or term that's defined in the Glossary in the back of the book.) It's attached to the bottom of a kind of long-lasting thunderstorm known as a **supercell**, which is characterized by an area of rotation known as a **mesocylone**, perhaps ten miles or so in diameter. Henson noted that the most obvious sign of a mesocyclone in the photo is the banded region with a corkscrew look above where the wall cloud meets the cloud above. Hurricanes are made of organized groups of smaller thunderstorms and do not contain supercells.

Henson knew that Mike Hollingshead, a storm chaser from Blair, Nebraska, shot the photo. According to Hollingshead's Web site, it was taken near Alvo, Nebraska, in the late afternoon of June 13, 2004, and soon after he took the photo, small tornadoes formed under the left side of the cloud. This brings us to the dangers from such clouds and how they differ from those posed by hurricanes.

Supercells can produce the strongest tornadoes. If you ever see a cloud that looks like this one, head for a place where you can find shelter from a tornado. As we discuss in Chapter 8, forecasters can predict where and when tornadoes are most likely to develop, but not exactly where and when one will hit. You have, at best, only minutes to take shelter from a tornado. With hurricanes, on the other hand, forecasters can almost always give you plenty of time to find a safe place.

As with almost every topic covered in this book, the Web site has more about this photo and links to other sites, including Hollingshead's. Use the "Comments and Questions" link on the site's home page to ask us about anything you can't find.

But first, join us in the chapters ahead as we unravel the mysteries of how the atmosphere and oceans produce the weather and climate that are a big part of our lives. We begin coverage of our weather topics in Chapter 1 by discussing science and the famous weather disasters that have caught the public's attention. In Chapter 2, we embark on a detailed exploration of the science of weather and oceanography, including how energy from the sun powers Earth's weather and how the atmosphere and oceans move energy around the Earth, creating its many different climates.

In Chapters 3 and 4, we cover the basic science of the forces that create winds, clouds, rain, snow, ice, and other aspects of weather. In Chapter 5, we put these pieces together to see how they create the global weather patterns that cause not only our daily weather but also the earth's overall climate, including changes in Earth's climate.

After seeing how weather is observed and forecast in Chapters 6 and 7, we turn to particular kinds of weather events, including thunderstorms and tornadoes in Chapter 8, middle-size weather systems such as clusters of thunderstorms in Chapter 9, hurricanes in Chapter 10, and some of the kinds of dangerous weather you seldom see on the evening news, such as dangerous heat, in Chapter 11.

Our exploration culminates in Chapter 12 with a look at the science related to one of the most important and often discussed issues of our time, a discussion without which this book would not be complete: how Earth's climate is changing.

—Jack Williams
Washington, D.C., Winter 2008

> *How to cope with disaster, live*
>
> *with the weather, and enjoy the*
>
> *sky's wonders and science*

CHAPTER 1

At 10:11 a.m., Sunday, August 28, 2005, Robert Ricks, a **meteorologist** at the National Weather Service (NWS) office in Slidell, Louisiana, sent an electronic bulletin reading: "Devastating damage expected…Hurricane Katrina…a most powerful **hurricane** with unprecedented strength…rivaling the intensity of Hurricane Camille of 1969. Most of the area will be uninhabitable for weeks…perhaps longer."

His bulletin warned: "Airborne debris will be widespread…and may include heavy items such as household appliances and even light vehicles. Sport utility vehicles and light trucks will be moved. The blown debris will create additional destruction. Persons…pets…and livestock exposed to the winds will face certain death if struck."

Ricks predicted the aftermath: "Power outages will last for weeks…water shortages will make human suffering incredible by modern standards."

Previous pages: A helicopter carrying Federal Emergency Management Agency urban search and rescue workers flies over flooded New Orleans five days after Hurricane Katrina hit Louisiana and Mississippi.

Ricks, who was born in New Orleans, knew his audience: "People who were on the fence, trying to make the decision to finally leave. I grew up in this area, I know people who never leave the city; their lives are confined to their neighborhood. It disturbs their comfort zone when they are asked to leave." In fact, he said of his father, who died in June 2005, "Had he still been alive for the storm, it is uncertain if he would have stayed, as [he had] in past storms."

During the forty-eight hours after Ricks sent his bulletin, Katrina moved inexorably inland with destructive power matched by few storms. The experiences of his extended family—uncles, aunts, cousins, their spouses, children, and grandchildren—mirror those of many of the more than a million men, women, and children in Louisiana, Mississippi, and Alabama. Ricks was fortunate in that none of his extended family members were among Katrina's estimated 1,600 fatalities.

Evacuation plans that Ricks and his immediate family made before Katrina threatened New Orleans ensured that his wife, Cynthia; their teenage son and daughter, Joshua and Lauren; and their miniature dachshund, Cocoa, would spend August 29 with relatives in Church Point, Louisiana. There they experienced generally sunny skies as Katrina wrecked New Orleans 100 miles to the east.

One of his aunts, Teresa Ricks, evacuated her home in Waveland, Mississippi before the storm. When Katrina's **eye**, with the storm's strongest winds swirling around it, came ashore at Waveland, the storm surge washed the house off its foundation, destroying it. Ricks' stepmother, Cathy Ricks, fled before the storm with her two sons and other family members to Millington, Tennessee, outside Memphis. Katrina destroyed her home in the New Orleans suburb of Poydras.

Other relatives decided to stay despite Ricks' calls urging them to flee. One of his aunts, Sylvia Guerin, argued that she had to stay to open her restaurant, Pudgy's Stuffed Potatoes in Chalmette, after the storm. She was among twenty relatives who stayed together at the two-story house of an uncle.

By the way, the restaurant that Ricks' Aunt Sylvia stayed behind to open the next morning blew up from a gas leak during the hurricane.

According to Ricks, "Shortly before Katrina's strongest winds hit, they were outside thinking, 'Man, is this all it's going to be?' Then they heard a roar and looked up the street to see a wall of water. They got inside and tried to shore up the door. Someone saw a wall bulging in. The house started to flood. I had told them to expect overtopping of levees, and they were well prepared with a boat." They shuttled everyone to a Mississippi River levee, about three blocks away, to ride out the storm.

Ricks says overtopping or failure of three levees allowed water from three directions to advance toward his uncle's house, putting it under fourteen feet of water. But the Mississippi River levee, in the area where the family found refuge, held. One thing that Ricks and his relatives know (as do most people who grow up in a flood-prone area) is that "if there's a flood, go to the levees. It's the high ground."

A success and a failure

It's hard to imagine that by Sunday morning, August 28, New Orleans had more than a few residents or visitors who didn't know that a Category 5 hurricane, with winds up to 160 mph and perhaps 20 feet of storm surge, was headed for them. The surge was bound to cause flooding when it washed into the marshes and waterways between the city and the Gulf of Mexico and into Lake Pontchartrain at the city's northern boundary. The National Hurricane Center had issued a hurricane watch Saturday morning that included metropoli-

tan New Orleans. Hurricane Center forecasts every six hours since then had made it clear that Katrina would hit as a major storm with winds faster than 111 mph. In Chapter 10, we look at how meteorologists made these forecasts, which were extraordinarily accurate.

For decades, hurricane scientists and forecasters had been predicting a weather disaster for New Orleans that could match or exceed the 1900 Galveston, Texas, hurricane, which killed at least 8,000 people. Newspapers, including the *Times-Picayune* in New Orleans; magazines; television programs; and books had all described what could happen when—not if—a major hurricane hit New Orleans. As Ricks told NBC's Brian Williams on September 15, 2005, "We always prepare for the big one; we just didn't think it was going to come this soon."

Despite the knowledge of the potential for such a hurricane, and forecasts that gave more than two days' warning, Katrina turned out to be one of the deadliest natural disasters in U.S. history.

Katrina could have been much worse. Years of news stories about the consequences of a major hurricane and the Katrina warnings, including the one Ricks sent out that Sunday morning, prompted an estimated 85 percent of the approximately 1.2 million people in the storm's target areas to leave. Previous surveys and studies had concluded that perhaps 50 percent of New Orleans residents would leave. We have no way of knowing how many evacuees would have died had they not fled. Photos of the **flood** water and damage in New Orleans, and of entire Mississippi communities washed away by Katrina's storm surge, offer strong evidence that thousands more would have died had they stayed.

In the days after Katrina, the news media's focus wasn't on the success of prompting thousands to evacuate, but on the hundreds trapped in New Orleans. The paradox of Katrina is that the response both succeeded marvelously and failed miserably. Obviously, accurate weather forecasts aren't enough. Perhaps even more disturbing is that coastal residents can't always count on as much warning as the National Hurricane Center gave for Katrina, especially for a major hurricane. The 1935 Labor Day hurricane was the strongest ever to make landfall in the United States, and it haunts hurricane forecasters and emergency managers to this day. It grew from a **tropical storm** with winds slower than 65 mph in the Bahamas to a Category 5 hurricane hitting the Florida Keys in less than 48 hours.

Katrina intensified the debate in the United States about how lives can be saved and damage can be reduced in future weather catastrophes. As Jeff Rosenfeld, the editor in chief of the *Bulletin of the American Meteorological Society (BAMS)*, wrote in the November 2005 issue, "The death toll from Katrina…was indeed closer to that of 9/11 than any homeland disaster in the last 60 years. And the costs from Katrina will be several times that of the terrorist attacks."

Rosenfeld argued that, "For too many of us, a Katrina was too remote, too incomprehensible, or simply too gargantuan. A near-miss bombing in lower Manhattan in 1993 and mounting intelligence did not lead to enough prevention to avoid the destruction of the World Trade Center. Neither did the near miss of Hurricane Georges [in 1998] or countless scientific studies about every imaginable aspect of New Orleans' vulnerability save enough lives or prevent over $100 billion damage on the Gulf Coast."

Living with weather

Hurricanes are just one of the many weather and climate issues that nations and societies around the world will face in the coming years and decades. In this chapter and in the following chapters, we examine how weather and climate work and some of the things you can do to live more comfortably and safely with the weather.

Good sense says you should know about weather dangers whether you live along the U.S. Gulf of Mexico or Atlantic coasts, where a hurricane could hit; in the Southeast or Great Plains, where fierce tornadoes are most likely; or on a California hillside, which winter rain could turn into a slurry of sliding mud, rocks, and debris.

During television interviews in Katrina's aftermath, some residents compared the dangers of living near the Gulf of Mexico with the **tornado** danger on the Great Plains. Such comments reflect a stunning lack of comprehension of the relative dangers. Even the worst tornado outbreak cannot lay waste to hundreds of square miles the way that Katrina and many smaller and weaker hurricanes have done.

Most places have their particular weather dangers. Floods kill people in areas that are normally wet as well as those that are normally dry. Lightning is a danger anywhere you hear thunder.

While dramatic storms might be frightening, weather in general need not be feared. As with many things that inspire fear, knowledge can be the antidote, especially if you use that knowledge wisely. Ricks, for instance, used his meteorological knowledge and bought a house high enough and far enough away from the water to avoid storm-surge flooding. It's on the north side of Lake Pontchartrain, across the lake from New Orleans and 39 feet above sea level. "I have a love for the water," he says, "and would have loved to live closer to the lake," but he didn't want to worry about flooding. He was worried, however, about trees falling on the house. Ricks says his family doesn't regret evacuating even though 100 mph winds and a nearby tornado did no serious damage to the house; staying would have been frightening.

His family's evacuation plan exemplifies the preparations that anyone who lives along the U.S. Gulf or Atlantic coasts should make before the hurricane season begins each year. The plan had options of going east, north, or west depending on where Katrina seemed likely to head. They had a reservation at a hotel in Biloxi, Mississippi, if east were the way to go, and arrangements to stay with a Weather Service colleague near Jackson, Mississippi, if north were the safest direction. Based on the forecast, they decided that the third option, heading west to stay with relatives in San Antonio, Texas, was the best choice. Cynthia Ricks and their children left on Sunday morning a couple of hours before her husband sent the bulletin. He had gone to work at 4 a.m. at the Slidell NWS office, forty miles northeast of downtown New Orleans on the north side of Lake Pontchartrain, for the twelve-hour shift that everyone there works during emergencies. Ricks brought with him five days' worth of food and clothing because everyone at the office knew he or she could be stuck there for days in Katrina's aftermath.

Even though Cynthia Ricks and the children didn't wait until the last minute to leave, evacuees were clogging highways heading west. What should have been a three-hour drive to the area of Church Point, Louisiana, took nine hours. With Church Point unlikely to be in Katrina's path,

Cynthia Ricks elected to stay with relatives there rather than risking another twelve or more hours in traffic trying to reach San Antonio.

More than disasters

Fortunately for most of the world, especially the more populated areas, the weather is good much more often than bad, even considering disagreements about what constitutes good and bad weather. Think back on your life and how often you had to take special precautions (or wish you had) because of weather. Unless you've gone looking for trouble, days when the weather required precautions were almost certainly rare.

Most bad weather days aren't caused by extreme events, such as Hurricane Katrina, but are what Roger Pielke Jr. of the Cooperative Institute for Research in Environmental Sciences (CIRES) and Richard Carbone of the National Center for Atmospheric Research (NCAR), both in Boulder, Colorado, refer to as "routinely disruptive weather." In their article, "Weather Impacts, Forecasts, and Policy" in the March 2002 issue of *BAMS*, they define such weather as "not extreme, but significant enough to warrant behavioral adjustments." Such events could include snow that requires you to shovel your driveway before driving to work on roads that have been plowed.

If the **atmosphere** is behaving the way most people would like, and you've decided how to dress to go outdoors today, you might be finished thinking about the weather for that day. In fact, maybe you think that only meteorologists and a few hobbyists with backyard weather instruments really care about the weather.

That's not the case, however, because weather, good and bad, is a part of the economy. Exactly how big a part is an open question because, as Pielke and Carbone write in their 2002 article, "There is no centralized collection of data and no standardized methodology" for assessing weather's economic effects.

Even though the costs of weather and benefits of forecasts are hard to pin down, economists and others try to do just that. For instance, The National Oceanic and Atmospheric Administration (NOAA) says in its 2006 report *Economic Statistics for NOAA* that industries affected the most by weather and climate account for about one-third of the nation's

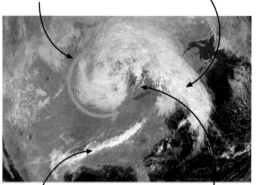

Satellites give the big picture

This weather satellite image shows a storm that formed over the western Plains and moved into the Midwest with high winds and tornadoes. The worst was over at the time of this image, 11 a.m. on April 12, 2001. The image shows that clouds over most of the eastern United States are part of the same system—Chapter 5 explains how this works.

Clouds show counterclockwise swirl of winds around the storm's center over Lake Superior.

Image shows clouds aren't likely to clear soon over Albany, New York.

Remnants of a line of thunderstorms that caused damage during the previous two days move toward the east.

The sky is clearing over Detroit, but clouds to the southwest could arrive in a few hours.

In this book, capitalized "Weather Service" and "NWS" refer to the U.S. National Weather Service.

gross domestic product, or $4 trillion in 2005 dollars. Such industries include finance, insurance, real estate, retail and wholesale trade and manufacturing, agriculture, construction, energy distribution, and outdoor recreation.

If you're a regular air traveler you're sure to have experienced weather effects on aviation. The Air Transport Association reports that air-traffic delays cost the airlines $6 billion a year, with weather causing 70 percent of these delays. Major disasters such as Katrina, which closed airports in New Orleans and some nearby places for weeks, add to aviation's weather costs. But routine weather, such as small **thunderstorms**, fog, and low clouds that cause flight delays and cancellations, are much more frequent than major disasters, and their total cost is higher than the costs of disasters to individual airlines. Weather also adds to the costs of traveling or shipping on highways and railroads. NOAA says clearing snow from streets, roads, and highways costs $2 billion a year in the United States. A big Northeastern U.S. snowstorm that shuts down cities from Washington, D.C., north to Boston, can cost $10 billion a day in lost retail business, wages, and tax revenues.

A gallery of clouds

Clouds form and stay in the sky only when the air is rising—sometimes only inches per hour, at other times at 100 mph. In Chapter 4, we see how clouds form and sometimes produce rain, snow, or ice and why some clouds are flat stratus clouds, while others are piled-up cumulus clouds.

The colors of clouds

Cloud drops less than 20 micrometers in diameter scatter all wavelengths of white sunlight in all directions.

As drops grow larger than 20 millimeters—as they do before rain, snow, or ice falls—they begin absorbing a little sunlight, but they also scatter more light than they absorb.

Half of the light reaching a cloud less than 3,000 feet thick goes through the cloud while the other half is scattered, making the cloud top white and the bottom a light gray.

Clouds in the shadows of other clouds are also gray.

Hardly any sunlight makes it through a cloud more than 3,000 feet thick.

1 Fall streaks are ice crystals that are being pushed by high-altitude winds as they fall from tiny, puffy clouds, which are hardly visible at the tops of the streaks. Fall streaks are often called mares' tails.

A patch of more-humid, unstable air causes the patch of thicker clouds and fall streaks.

A solid sheet of cirrostratus clouds is moving in from the north to make the next day overcast.

Perspective makes fair-weather cumulus clouds near the ground in the distance appear closer together than they really are.

2 Lenticular clouds like the ones below are commonly seen downwind from mountains. Wind blowing over the mountains from the right continues rising and then descends as shown by the arrow on the drawing.

When wind descends and warms, cloud drops evaporate.

Clouds form where air rises and cools enough for condensation to begin.

Wind carries cloud drops through the cloud. The drops are always moving, but the clouds we see don't move.

3 Cumulus congestus clouds like these can grow into large cumulonimbus (thunderhead) clouds.

The cloud is evaporating where air is no longer rising.

A cauliflower appearance changes to a smoother, fibrous look when ice crystals form.

Air is rising and the cloud growing in cauliflower-like areas.

4 Fair-weather cumulus clouds like these don't produce rain or snow. Sometimes, however, clouds that look like fair-weather cumulus grow into larger, precipitation-producing clouds.

When air no longer is rising into it, a cloud begins evaporating.

Air is sinking in clear air between clouds.

A cloud's flat bottom is where the humidity in rising air begins condensing into cloud drops. A cloud's top is where air is no longer rising.

Low, flat stratus clouds often cover all or most of the sky. Fog is a stratus cloud on the ground.

How water drops make rainbows

You see a rainbow only when the sun is behind you, as you look toward water drops in the air. Light is refracted (slightly bent) as it enters a drop, reflected from the back of the drop (sometimes more than once) and refracted again as it leaves the drop.

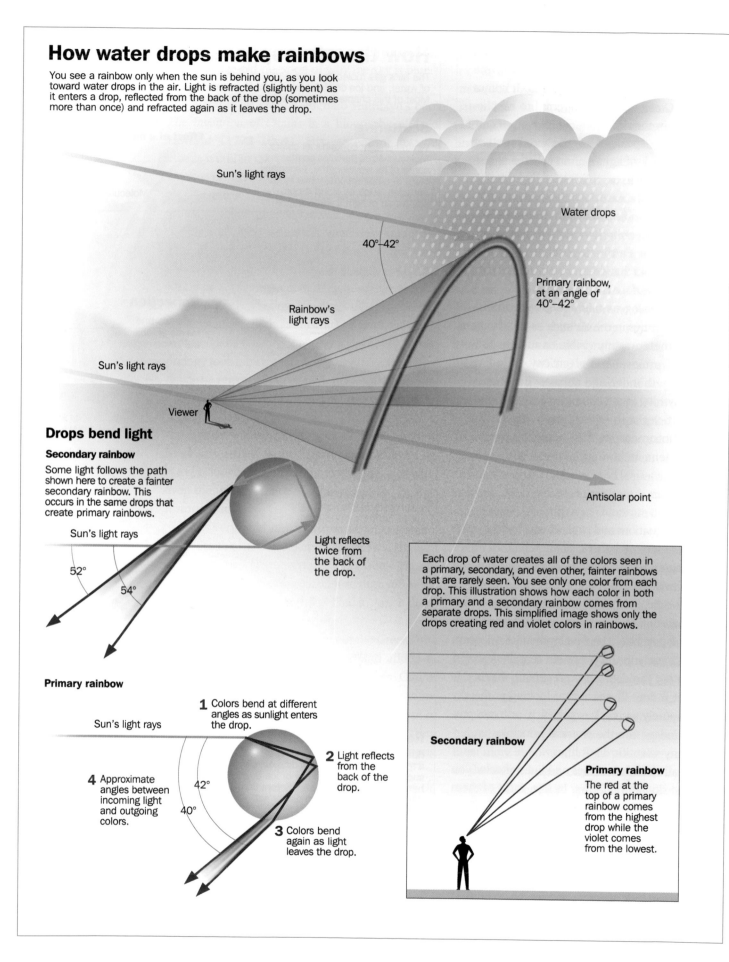

Sun's light rays

Water drops

40°–42°

Primary rainbow, at an angle of 40°–42°

Rainbow's light rays

Sun's light rays

Viewer

Antisolar point

Drops bend light

Secondary rainbow

Some light follows the path shown here to create a fainter secondary rainbow. This occurs in the same drops that create primary rainbows.

Sun's light rays

52°

54°

Light reflects twice from the back of the drop.

Primary rainbow

Sun's light rays

1 Colors bend at different angles as sunlight enters the drop.

2 Light reflects from the back of the drop.

4 Approximate angles between incoming light and outgoing colors.

42°

40°

3 Colors bend again as light leaves the drop.

Each drop of water creates all of the colors seen in a primary, secondary, and even other, fainter rainbows that are rarely seen. You see only one color from each drop. This illustration shows how each color in both a primary and a secondary rainbow comes from separate drops. This simplified image shows only the drops creating red and violet colors in rainbows.

Secondary rainbow

Primary rainbow

The red at the top of a primary rainbow comes from the highest drop while the violet comes from the lowest.

fusing because it refers to both the sun's outer atmosphere, which can be seen during eclipses, and to a disc of light surrounding the moon or the sun. Light being diffracted as it passes through clouds causes coronas and iridescence. If a circle of colored light (the colors could be faint) looks like it is touching the moon or sun it is a corona. A corona is usually the ring around the moon that can be a sign of coming rain.

Omens and wonders in the sky. Once you learn a little about meteorology and start looking at the sky, you'll find you can amaze others by pointing out phenomena such as halos and sun dogs, which are caused by sunlight being refracted, or bent, by ice crystals.

Sun dogs—their more formal name is parhelia—on either side of the sun are the most common kind of halo. Across most of North America you might see at least one sun dog every few days when high, thin clouds cover at least part of the sky.

To see if a narrow streak of light circling or partly circling the sun is a 22-degree halo or a splotch of light is a sun dog, extend your arm and spread your fingers. If your thumb is over the sun, the tip of your little finger will touch or be very close to a 22-degree halo or a sun dog.

When you see a sun dog or sun dogs and the sun is about 15 to 25 degrees above the horizon, look straight up. If the cloud causing the sun dog stretches to the zenith—the sky directly above you—you have a good chance of seeing a circumzenith arc as well, as Warren Tape and Jarmo Moilanen say in their book *Atmospheric Halos and the Search for Angle X*. If you watch the sky, you have a good chance of seeing a circumzenith arc maybe twenty-five times a year.

On rare occasions, conditions are right for displays of several kinds of halos at the same time. In the distant past, when people saw anything out of the ordinary in the sky, such as a comet or a display of several halos, they often interpreted the phenomenon as omens. Such events could cause consternation. In their book, Tape and Moilanen show a seventeenth-century engraving of a halo display over Nuremberg, Germany, on April 19, 1630. The engraving's text warns: "God threatens through word and deed, and God threatens through Nature."

While many in the seventeenth century likely felt uneasy at the sight of halos in the sky, a few were seeking what caused them, including Descartes. In 1637, Descartes suggested that giant rings of ice in the sky caused sun dogs. While we know now that Descartes was wrong, we have to realize that neither he nor anyone else in his time had any way of knowing that rings of ice do not form in the sky or that cirrus clouds are made of tiny ice crystals that cause halos. In 1662, the Dutch mathematician and astronomer Christian Huygens (1629–1695) worked out a theory of halos. Using the geometry of how light rays could be bent, he showed that transparent cylinders of water or ice with opaque cores could create a display that looked somewhat like a sun dog. He extended his idea to show how such cylinders could cause other halos.

Here the halo science story takes a turn that shows science is far from being as straightforward as you might think from reading science textbooks. In 1681, the French scientist Edme Mariotte (1620–1684) proposed that small ice crystals, which act like prisms, could bend light to cause halos. He even showed how a particular kind of crystal could form a 22-degree halo and, when turned in a different direction, could create sun dogs.

Today we know Mariotte was on the right track, but he was ignored until early in the nineteenth century because no one had any way of testing the differing hypotheses of Huygens and Mariotte. Both made fairly good predictions for 22-degree halos and sun dogs, which were the only kinds of halos scientists considered. However, at the beginning of the nineteenth century when scientists began trying to explain less-common kinds of

It's time to retire Roy G. Biv
In addition to their scientific and cultural interest, rainbows are also a prime example of how students, even in primary school, should develop a scientific attitude that asks "is this true?" when what they see doesn't agree with what they are told.

For years, American school children have been told that the name "Roy G. Biv" will help them remember the colors of the rainbow. If you search the Web for "rainbow colors" you'll find sites, including some purporting to be educational, selling the idea of Roy G. Biv for the colors red, orange, yellow, green, blue, indigo, and violet as the "colors of the rainbow."

But anyone who looks at a real rainbow will tell you that Aristotle's idea that the rainbow has three colors comes closer to being the case than Roy G. Biv. Lee and Fraser write in *The Rainbow Bridge*, "The number seven derives not strictly from visual observation but also from Newton's belief that sight and hearing are related. Because each musical octave contains seven tones and semitones, he reasoned that the spectrum should have seven colors."

Through the centuries, artists have depicted rainbows in myriad ways that offer a window on the artist's culture. For an interesting look at this topic, see Lee and Fraser, *The Rainbow Bridge: Rainbows in Art, Myth, and Science.*

Ice crystals create halos

Flat ice crystals falling with their hexagonal faces down form the sun dogs on both sides of the sun and also the circumzenith arc overhead in the photo. Crystals with different shapes and orientations create other kinds of halos, including the 22-degree halo around the sun and the tangent arc.

Because of space constraints, the sun dogs and circumzenith arc below are drawn closer to the sun than in the photo. Forward scattering of sunlight by tiny particles in the air creates a solar aureole that makes the sun look larger in the photo than it would be in perfectly clear air.

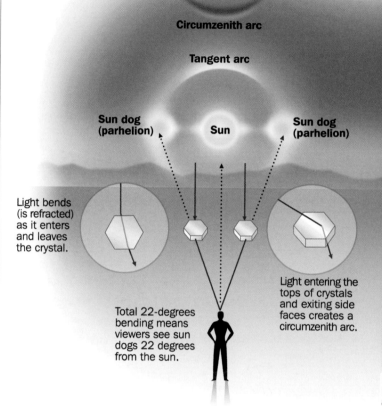

Circumzenith arc

Tangent arc

Sun dog (parhelion)   **Sun**   **Sun dog (parhelion)**

Light bends (is refracted) as it enters and leaves the crystal.

Total 22-degrees bending means viewers see sun dogs 22 degrees from the sun.

Light entering the tops of crystals and exiting side faces creates a circumzenith arc.

Ken Tape takes a picture of his father, Walter, observing a halo in northern Alaska; Ken used his father's head to block the sun and capure the display on camera.

halos, they recognized the superiority of Mariotte's hypothesis. The evidence for Mariotte's hypothesis has continued to grow. For instance, scientists now make direct observations of high clouds by going up in airplanes to capture ice crystals, which they examine under microscopes. (We will read about one of these scientists in Chapter 4.)

Generally, when you see a 22-degree halo or a sun dog, the ice crystals responsible are more than 20,000 feet up. If you live, or have ever lived in a place with cold winters, you've surely seen evidence that ice crystals that create halos can float in the air very near the ground.

One of the kinds of halos you sometimes see high in the sky is a sun pillar, which looks like a streak of light coming down from the sun toward the earth or, if the sun is low in the sky, a streak shooting up from the sun. On a still night when the

temperature is well below freezing, the air can be filled with tiny ice crystals known as diamond dust. On such nights, you sometimes see narrow beams of light going up from the uncovered tops of lights such as streetlights.

To see really spectacular halo displays, you should head for the polar regions, to places such as the U.S. Amundsen-Scott Base at the South Pole, where the highest temperature ever recorded is 8 degrees above zero Fahrenheit. Tape notes that while "the polar regions tend to have better halos than elsewhere, nobody really knows why. Cold by itself is not an explanation since high clouds in **temperate climates** are plenty cold."

Unlike the men and women of seventeenth-century Europe who feared that God was sending warnings, the men and women at the South Pole on January 11, 1999, relished the spectacle when

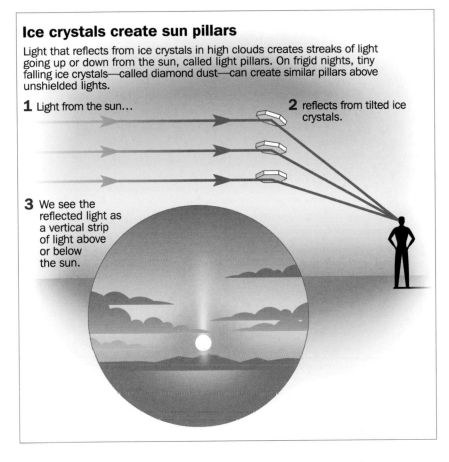

Ice crystals create sun pillars

Light that reflects from ice crystals in high clouds creates streaks of light going up or down from the sun, called light pillars. On frigid nights, tiny falling ice crystals—called diamond dust—can create similar pillars above unshielded lights.

1 Light from the sun…

2 reflects from tilted ice crystals.

3 We see the reflected light as a vertical strip of light above or below the sun.

ice crystals is nowadays pretty much computer business. You take photos of the display and simulate it by experimenting with differently shaped crystals in different orientations until you get a result that matches the display. Taking ice crystal samples can further confirm that the crystal shapes used in simulations are roughly correct."

Scientists in the nineteenth and early twentieth centuries worked out the mathematics of the shapes and orientations of ice crystals needed with the sun at various altitudes above the horizon to form certain kinds of halos. But many of the formulas were too complex to easily calculate until scientists started using computers. The fact that computers can churn out different representations of halos for various crystals and sun angles doesn't mean, however, that discoveries aren't waiting to be made. Tape and Moilanen say that researchers still have much to learn. "We have no doubt," they write, "that with increased awareness new halos will be seen and photographed." They say that "anyone who has modest camera equipment and is alert for what to watch for can make a contribution." Their book includes a guide to halo photography.

Why is the sky blue? Scientists know now that molecular scattering of blue light turns the sky blue, but this knowledge was not easily gained. The British physicist John William Strutt (1842–1919), who was the Third Baron Rayleigh and is usually referred to as Lord Rayleigh, developed the theory of electromagnetic scattering in the 1870s. To scatter **electromagnetic radiation**, particles have to be much smaller than the wavelength of the radiation. The amount of scattering depends on both the wavelength of the radiation, such as light, and the

In common use a halo is circular in shape, but scientists who study atmospheric optical phenomena have extended the term to refer to all photometeors, including rings, arcs, pillars, or bright spots, around the sun or moon that are caused by clouds made of ice crystals or by ice crystals floating in the air.

at least twenty-two different forms of halo graced the sky during a display that lasted fifty minutes.

"We knew we were witnessing something extraordinary. Of course, every moment at the bottom of the earth is extraordinary, but this was more exciting than usual," says Mary E. Hanson, a National Science Foundation public affairs officer.

"It was one of those crispy-clear days, and the sky and ground both seemed sprinkled with diamonds. Then these bands of translucent shimmer started to appear in the sky…lots of them. If you were a religious person, you might have thought they were heavenly beings or angels. We didn't know what they were until the Finnish scientists told us. They were as excited as the typically reticent Finns are likely to get. (I'm a Finnish-American so I can say that.) That's how I knew it was a true scientific phenomenon."

Jarmo Moilanen (co-author of *Atmospheric Halos*), Marko Riikonen, and Marko Pekkola were the three Finnish scientists studying halos at the South Pole that year. In addition to taking hundreds of photographs of the displays they saw, the Finns captured crystals from the air to study under a microscope. Riikonen explains, "relating the halos and

The sky does not contain anything that's blue; it's blue because the molecules of nitrogen, oxygen, and other gases in the air are the right size to scatter (reflect) violet and blue light.

Astronauts see a black sky and a white sun above about 20 miles.

Sun's light contains all colors.

3 Approximately 18 miles up, there are enough air molecules to begin scattering blue light.

5 When the sun is high in the sky, other colors continue mostly unimpeded to the surface.

6 As light travels through the air blue, then green are completely scattered, leaving yellow, orange, and red.

4 Molecules scatter violet and blue light in all directions.

7 Further scattering removes yellow and orange, leaving red.

8 Red is first to appear at sunrise, last to fade at sunset.

size of the particles. As it turns out, the molecules of nitrogen and oxygen, which account for 99 percent of the air's molecules, are more efficient for scattering the wavelength of blue light than the wavelengths of other colors.

The discovery of Rayleigh scattering explained why the sky is blue, a question that scientists had been trying to answer since the time of the Greeks with no success. Previous ideas based on the scattering of light by water drops, water vapor, or dust particles in the air did not stand up to scientific scrutiny.

Even more importantly, the discovery that the sky is blue because the molecules of the air scatter blue light more efficiently than other wavelengths

helped confirm the hypothesis that all matter is made of extremely tiny atoms and molecules (which are made of atoms). Rayleigh's theory also enabled scientists to calculate the size of the air's molecules (since earlier experiments had established the wavelengths of different colors of light). The idea of atoms had been around since the time of the Greeks, but nineteenth-century scientists were making an increasingly compelling case for it. For instance, Maxwell, whose electromagnetic theory we discussed above in relation to rainbows, also helped develop the theory that the heat we feel results from the motion of molecules.

Maxwell's suggestion to Rayleigh that scattering by the molecules of the air could explain why the sky is blue, led him to consider the problem over several years and eventually do the calculations showing that molecules are the right size to scatter blue light, making the sky blue. Over the years, laboratory experiments and other evidence have confirmed this. A blue sky is more than a blue sky; it helps tell the story of how scientists go about understanding the physical world.

Summary and looking ahead

If you can't imagine yourself becoming a scientist or meteorologist, you are likely a daily consumer of weather information. This information could be as simple as what you see when you look out the window before leaving home each day. Most people also obtain weather information from radio, television, the Internet, or a newspaper.

Many people preparing to make a major purchase, such as a new car or house, or an investment decision, first research the topic, often at great length. They want to become educated consumers. Becoming an educated consumer of weather information isn't likely to save you a lot of money, but it can make your life easier and could even help you avoid disaster. This book will guide your efforts to become an educated weather consumer.

As an educated weather consumer, you'll be able to make better use of weather forecasts, because you'll know what they can and can't tell you. For instance, reliable, detailed predictions of which days will be wet and which will be dry more than a week or so in advance aren't possible.

Another important aspect of using weather forecasts is that you often have to make decisions, perhaps life and death decisions, based on probabilities, not categorical facts such as "the winds ARE going to blow faster than 100 mph tomorrow." The Hurricane Katrina forecasts for New Orleans are a perfect example of how your responses to forecasts have to be based on probabilities, not a firm statement that 100 mph winds will hit your neighborhood tomorrow. At 10 a.m. on Saturday, August 27, 2005, when the Hurricane Center issued the hurricane watch that included New Orleans, the odds were only 19 percent that Katrina's eye would pass within seventy-five miles of the city.

When Ricks issued his bulletin at 10 a.m. on Sunday (as described at the beginning of this chapter), the odds of Katrina's center passing within seventy-five miles of New Orleans had risen to "only" 35 percent.

Landfall of a major hurricane over a particular area on a particular day is an extremely rare event. For the sake of argument, let's say that such an event occurs every 50 years during the hurricane season at a specific location vulnerable to hurricanes. Given that the hurricane season is roughly 150 days, the observed probability for the event would be approximately 1 in 7,500, or 0.013 percent. When the forecast probability (in this case 35 percent) vastly exceeds the observed probability of a rare and potentially catastrophic event, be prepared to take cover.

Next we will begin our detailed exploration of the science of weather and oceanography by discussing, in Chapter 2, how energy from the sun powers Earth's weather and how the atmosphere and oceans move energy around the earth, creating its many different climates.

At meteorology's center

No history of the atmospheric sciences since World War II is complete without the stories of Joanne and Bob Simpson.

When forecasters said Katrina was a "Category 5" hurricane, the number referred to the scale that Herbert Saffir created in the 1960s and Bob Simpson modified in the early 1970s when he was di-

Bob and Joanne Simpson at the Roosevelt Roads Naval Air Station in Puerto Rico in 1964 after a Project Stormfury flight in the Navy WC-121 Super Constellation behind them.

rector of the National Hurricane Center (NHC). Research flights that scientists made into Katrina were carrying on work he had initiated in 1945.

In Chapter 5, we will see how the Hadley circulation in the tropics helps drive the atmosphere's global circulation. Today's meteorologists consider the work of Joanne Simpson and Herbert Riehl, published in 1958, the basis of our current understanding of the Hadley circulation.

In addition to being the first scientist to conduct hurricane research from airplanes, Bob proposed, organized, and initially ran the National Hurricane Research Project, which began in 1956. It is the forerunner of today's NOAA Hurricane Research Division. In 1964, when he was the Weather Bureau's deputy director of research, Bob established the National Severe Storms Laboratory in Norman, Oklahoma, which among its many accomplishments developed Doppler weather radar.

Bob began his meteorological career in 1939

with a $1,440-a-year, entry-level job at the U.S. Weather Bureau but quickly advanced, working as a forecaster in New Orleans and Miami and studying at the University of Chicago. In 1945 the Bureau sent him to Panama to help the Army Air Forces establish a tropical weather forecasting school. While there, he organized and flew on the first hurricane research flight.

Joanne began her meteorological career after earning her bachelor's degree at the University of Chicago in 1943. She taught weather to aviation cadets at the University of Chicago and New York University while completing her master's degree in meteorology at Chicago. When World War II ended, she, like other women who had been doing "men's work," was expected to get married and settle down as a housewife. But, "I didn't want to go home and mop the floor." Instead, in 1947 she became the first woman in the world to earn a PhD in meteorology. Her dissertation was about tropical clouds, a topic then considered outside the meteorological mainstream.

At the time, Bob was one of the few other scientists interested in tropical weather. He headed Bureau operations in Hawaii and the Pacific from 1948 until 1952 and was assigned to the Bureau's Washington, D.C., Headquarters before and after that. When he could get away from his other duties, he hitched rides on military hurricane reconnaissance flights. These convinced him that scientists needed focused research flights to learn how hurricanes work.

Neither the Eisenhower administration nor Congress saw a need for hurricane research until hurricanes Carol, Edna, and Hazel hit the East Coast in 1954, doing at least $750 million (in 1954 dollars) in damage and killing more than 150 people. The resulting outcry led Congress to fund the National Hurricane Research Project, which the Weather Bureau appointed Bob to organize and run.

The Bureau asked leading atmospheric scientists to advise and participate, including Riehl and Joanne. By the mid-1950s, Joanne had become internationally known for her work on tropical weather and clouds.

More than three decades later, Bob said that this was the beginning of "a lifelong, close friendship with Riehl." It was also "the beginning of my scientific association and collaboration with Joanne." This "melded into a personal relationship culminating in our marriage in January 1965 and the beginning of a long, happy, and fruitful life together."

The Simpsons were not only at the center of much important atmospheric science but also some of the bitter bureaucratic and political battles that can accompany weather and climate research and prediction.

For example, in 1969 Bob feared that he and some of his superiors would be fired after he gave Vice President Spiro Agnew his forthright opinion following an aerial tour on Air Force One of Hurricane Camille's damage. Simpson told the Vice President that the administration's refusal to allocate funds to upgrade military hurricane hunter airplanes had "placed the forecasters and the warning systems in great jeopardy." Without good airplane data, Bob told Agnew, NHC forecasters were hindered even though they correctly forecasted that Camille would be what is now called a Category 5 storm when it hit Mississippi. Civil defense officials credited that forecast with saving hundreds of lives by prompting last-minute evacuations. But Bob feared forecasters might miss the next such vital forecast without good aircraft data. Agnew's report prompted President Nixon to "chew out" weather and military officials, Bob says, but no one was fired and the airplanes were upgraded.

Some of the bitterest battles were over **cloud seeding**. In 1961, Bob had started Project Stormfury to investigate the hypothesis, which he and Joanne had developed, that seeding the largest clouds around a hurricane's eye could weaken a storm. In the early 1960s, Joanne arranged to take part in Stormfury because, among other things, it offered a way to test the computer **model** of clouds she had written—the world's first such model. (Cloud seeding is a technique using silver iodide or other substances to produce or enhance snow or rainfall from clouds.)

From 1964 to 1974, Joanne was director of the government's Experimental Meteorological Laboratory, which included Stormfury and later the Florida Area Cumulus Experiment, which experimented with seeding clouds over Florida. She left because she no longer wanted to work "at a political interface of science with so much hassle and so much unpleasantness that it became impossible to do any work."

During the first part of her career, Joanne often worked with men who didn't think she or any other woman belonged in science. Even when she was accepted as a scientist, she had to cope with issues such as finding good childcare. In 1973, she wrote in a *New York Academy of Sciences Annals* article, "I think that the difficulties faced by a woman trying to combine top-level achievement with marriage and motherhood are close to prohibitive."

In 1979, Joanne became head of the severe storms branch at NASA's Goddard Space Flight Center in Greenbelt, Maryland. There, she discovered that for the first time in her career, "I could discuss science in the ladies room." Another big turning point came in 1983 when the American Meteorological Society presented her with the Carl-Gustav Rossby Research Medal, its highest honor.

She found the award especially meaningful because Rossby was one of her mentors. For instance, in 1955 he arranged for her to use his computer in Stockholm, Sweden—one of the few available anywhere to meteorologists then—to complete her groundbreaking scientific paper on a mathematical model of clouds. Joanne says that "although everybody thinks that Rossby was a theoretician, he actually also believed in being an observer and being a naturalist." Earning a pilot's license helped spark Joanne's interest in weather, and she combined theoretical work with observations, often from airplanes. "I wish people still had that kind of idea," she says, "because I think too may graduate students now are getting totally involved in models without exposing themselves to observations or data or the real atmosphere."

Looking back over a half century of meteorological research, Bob is convinced that anyone who's just beginning today can be assured that many unsolved problems await research. "If there is anything I've learned over the years, it's that with all we've learned, there's so much more we don't know. Every time we make a bit of progress it shows us something else we didn't know and have to address."

How the atmosphere and oceans use sunlight to create climates and make our planet livable

Tim Stanton's story is adapted, with permission, from The North Pole Was Here: Puzzles and Perils at the Top of the World, *by Andrew C. Revkin, published in 2006 by Kingfisher, a Houghton Mifflin Company imprint.*

Tim Stanton, an oceanographer from the Naval Postgraduate School in sunny, warm Monterey, California, was kneeling on the shifting **sea ice** cloaking the Arctic Ocean a few dozen miles from the North Pole. His gloves were off despite the 20-below-zero Fahrenheit chill as he tested a $40,000 instrument pack, which he called "my baby."

Stanton crammed the instruments in a buoy resembling an overgrown red and yellow Tootsie Roll Pop. He was preparing it to transmit data back to his lab via satellite, providing a view of the year-round distribution of heat in the water just beneath the ice and in the ice itself. He'd drilled a hole through the six-foot-thick floe and set the device down so that its shaft reached into the seawater below where it measured salinity, temperature, and the currents that can move heat up, down, or sideways.

Stanton studies how sea ice affects the flow of heat between water and air. While his project might seem arcane, it's a vital facet of the overall effort to understand what may happen to the Arctic as the earth's climate warms. For more than a century, scientists have been trying to figure out how the heat of the sunbaked tropics and the deep chill of the poles interact to shape the world's weather and ocean circulation. The influence of Arctic climate on conditions around the Northern Hemisphere, in fact, inspired the first integrated international earth science research effort when, in 1882–1883, eleven countries joined to conduct the first International Polar Year.

More recently, researchers have focused on the increasingly pressing question of how much polar ice may melt in response to the global buildup of **carbon dioxide** and other heat-trapping smokestack, tailpipe, and agricultural gases. Few places on a warming Earth are as important to understanding the climate as the Arctic. In most places, climate change is expected to be gradual, largely hidden by the normal variability of weather. Up here, however, a little global warming can result in a lot of smaller-scale, regional warming and trigger additional changes very quickly. One reason for this is that if the air here warms enough to melt the polar summer's blindingly brilliant-white sea ice,

Previous pages: One of the scientists involved in a study of heat and moisture exchanges in a tropical rain forest (the climatic opposite of the polar regions) photographed a downpour on the M'Beli River in the Republic of Congo.

the dark water it once covered will absorb energy that previously had been reflected back into space.

Satellites and supercomputers help provide clues, but much polar science still must be done on site. The only way to confirm a satellite estimate of, say, sea ice thickness, or to sharpen the uncertain picture of the future climate produced by a computer simulation, is to get into the field and consistently collect measurements of the ice, ocean, and atmosphere. Each April since 2000, a hardy team of oceanographers, climate scientists, and Arctic experts—Stanton among them for several years—has camped on the drifting sea ice near the North Pole to study how the flow of heat in the ocean below and atmosphere above is changing, both through natural and human influences. They are trying to take consistent measurements year after year at a place where nothing is consistent, where the ice on the Arctic Ocean is constantly shifting and drifting several hundred yards per hour, where the only permanent thing is the sea bed 14,000 feet below.

Scientific research at the Arctic end of the earth resembles an extreme sport. In contrast to the South Pole, where a frozen continent, while brutally harsh, provides a footing for a permanent base, the North Pole has no home base with spare parts, no mechanic on duty to fix something that breaks, and no firm ground to retreat to when things go sour.

It's a place where the sea ice under your tent can suddenly split open, revealing a black **lead** (a channel through floating ice) of steaming 28-degree ocean water; where help is at least 500 miles away; where even the simplest setback can threaten ambitious research and, potentially, the researchers themselves.

Surviving in this environment requires that a person possess the brainpower of a scientist and the brute strength of a furniture mover. Conducting research here demands the wile of a small-town me-

chanic, the determination (or recklessness) of an edge-testing athlete, and the courage to ward off polar bears with a shotgun.

Stanton designed his buoys to withstand the crushing pressures of colliding sea-ice floes that can raise house-high ridges of car-size blocks. Shortly after he installed a buoy one day, just such a ridge started to build under the parked Russian helicopters that were waiting for the science team. The ice beneath the team's feet vibrated and began chugging and huffing and shuddering like a freight train slowly leaving a station.

"There's no mechanical system you could really design to live through a thing like this," Stanton said, pointing as a heap of greenish ice slabs nearby rose a bit. "That's just the risk we take. [The buoy] could last two days, it could last two years—you just don't know."

Finally, there was no time for more tests; the Russian helicopter crew needed to return to a nearby tourist camp to take wealthy customers from Moscow to skydive at the North Pole.

Stanton walked away from his buoy, now a tiny dot in the crinkled whiteness, and prepared to head for home.

Weather, climate, and energy

In the previous section, we learned about the North Pole scientists who track one remote aspect of the global flow of energy. To understand weather and climate you need a complete picture of the global interplay of energy and matter in Earth's atmosphere (air), **hydrosphere** (the oceans and other water), **cryosphere** (ice), and **biosphere** (living things). Measuring and understanding energy flow is basic to all science, and it helps explain weather as well as provide an idea of what Earth's climate is likely to be a half century from now.

Above: Stanton and colleagues assemble his buoy at the North Pole.

Above left: Tim Stanton's "baby."

When water vapor evaporates into air that is much colder than the water, the water vapor begins condensing into tiny water droplets immediately above the water to form **steam fog**—it looks like steam coming from the water.

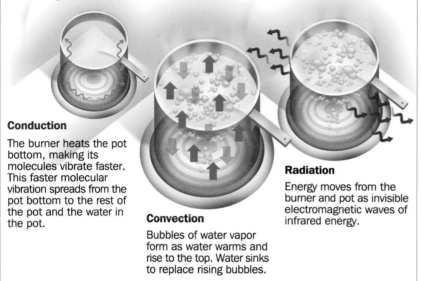

How heat travels

Heat from the sun supplies almost all of the energy that drives the weather. A pot of water on a stove illustrates the three ways that heat moves from the sun to Earth and through the atmosphere and oceans.

Conduction

The burner heats the pot bottom, making its molecules vibrate faster. This faster molecular vibration spreads from the pot bottom to the rest of the pot and the water in the pot.

Convection

Bubbles of water vapor form as water warms and rise to the top. Water sinks to replace rising bubbles.

Radiation

Energy moves from the burner and pot as invisible electromagnetic waves of infrared energy.

As Earth follows a nearly circular, elliptical orbit around the sun over the course of a year, its average distance from the sun is 93 million miles—a distance that electromagnetic radiation, including light, travels in eight minutes.

In this chapter, we examine the big picture of how the unequal distribution of energy from the sun powers Earth's weather and creates ocean currents. In succeeding chapters we look into how heat energy transforms into the mechanical energy of moving water and the world's many kinds of weather.

When water in a pot begins boiling, it is transporting heat via convection, which refers to the motion of a fluid (a gas or a liquid) that transports heat as it moves away from the heat's source. Unlike **convection**, heat moving through the air as electromagnetic radiation does not move the air, and heat moving from the bottom to the top of a pot on a stove via conduction does not move the water. **Convective** circulations are an important driver of weather. While the term convection can refer to any movement driven by density differences, meteorologists generally use it to refer to localized up and down movements of air, such as in thunderstorms. They use the term **advection** for the generally horizontal air movements that carry atmospheric properties, such as heat or humidity.

It starts with the sun. Astronomers describe the sun, which is large enough to hold a million Earths, as a rather ordinary star that has been producing energy for approximately 4.6 billion years. It should continue producing energy for another five billion years before expanding into a red giant star that engulfs its inner planets, including Earth. Nuclear fusion reactions that transform hydrogen atoms into helium atoms give off the huge amounts of energy that make stars, including our sun, glow over billions of years. Energy leaves the sun in all directions as electromagnetic radiation of several kinds, including gamma rays, x-rays, ultraviolet light, visible light (the only kind of electromagnetic energy we can see), infrared energy, and radio waves.

All objects give off electromagnetic radiation with the frequency, and thus the type of radiation, dependant on the object's temperature. You see this when you grill meat over hot coals. When you light the fire that ignites the coals, you see a flame with blue, yellow, and red colors. The blue part of the flame is the hottest, in the neighborhood of 2,500 degrees Fahrenheit, while the red part of the flame and the red glow from the charcoal as it begins burning are cooler, approximately 1,400 degrees. Later, after you've finished grilling and the charcoal is cooling, the red glow becomes dimmer and eventually disappears, but you can still feel the infrared heat energy radiating from the coals.

Most of the energy the sun radiates comes from its surface, which has a temperature of approximately 10,500 degrees Fahrenheit. About 44 percent of the sun's emitted energy is in the visual part of the spectrum and about 48 percent is infrared energy. Ultraviolet energy accounts for approximately 7 percent, with other wavelengths making up the remaining one or so percent. The small amount of ultraviolet energy from the sun that reaches the earth's surface causes people to sunburn and can damage living things in other ways. Fortunately, **ozone** in the upper atmosphere blocks much of the sun's ultraviolet energy.

The climatic balancing act. The total balance between incoming and outgoing radiative energy determines any planet's climate. To see what this means, let's begin by imagining what Earth would be like if it had no atmosphere or oceans. Such an Earth would have a daily range of temperatures much like that on Earth's moon, which has no ocean and an extremely thin atmosphere (some scientists don't like calling it an atmosphere). During the day, the sun shining on nothing but bare rock would heat the rocks up to approximately 200 de-

Electromagnetic waves

Light that we see (visible light), radio waves, infrared (heat) radiation, ultraviolet rays that cause sunburn, x-rays, and the nuclear radiation known as gamma rays are all electromagnetic waves that carry radiative energy. They are the same "kind" of waves, differing only in their frequency.

Wavelength and frequency

All waves have regular motion and carry energy as they move. A wave's frequency, which is how many waves pass a point in a given time such as a second, depends on the length of the waves (wavelength) and their speed, as shown here.

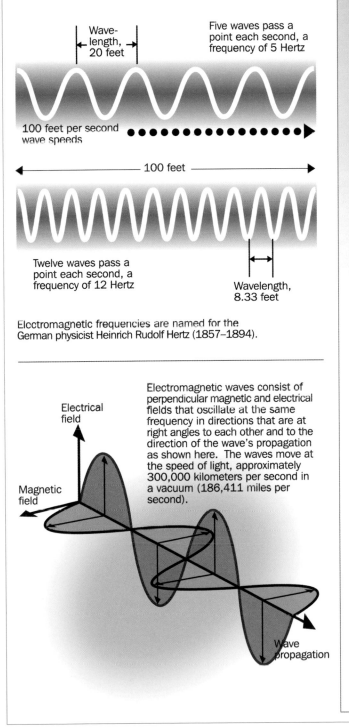

Wave-length, 20 feet

Five waves pass a point each second, a frequency of 5 Hertz

100 feet per second wave speeds

100 feet

Twelve waves pass a point each second, a frequency of 12 Hertz

Wavelength, 8.33 feet

Electromagnetic frequencies are named for the German physicist Heinrich Rudolf Hertz (1857–1894).

Electromagnetic waves consist of perpendicular magnetic and electrical fields that oscillate at the same frequency in directions that are at right angles to each other and to the direction of the wave's propagation as shown here. The waves move at the speed of light, approximately 300,000 kilometers per second in a vacuum (186,411 miles per second).

Electrical field

Magnetic field

Wave propagation

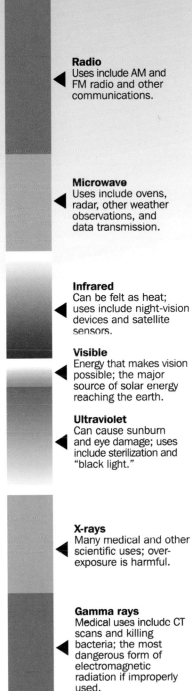

The electromagnetic spectrum

The different kinds of electromagnetic energy fade into one another. As you go down this image, the wavelengths become shorter and thus the frequencies become higher. Energy also increases as you go down, which is why ultraviolet energy, x-rays, and gamma rays are potentially more harmful to living things. A few uses or characteristics of each region are listed below.

Radio
Uses include AM and FM radio and other communications.

Microwave
Uses include ovens, radar, other weather observations, and data transmission.

Infrared
Can be felt as heat; uses include night-vision devices and satellite sensors.

Visible
Energy that makes vision possible; the major source of solar energy reaching the earth.

Ultraviolet
Can cause sunburn and eye damage; uses include sterilization and "black light."

X-rays
Many medical and other scientific uses; overexposure is harmful.

Gamma rays
Medical uses include CT scans and killing bacteria; the most dangerous form of electromagnetic radiation if improperly used.

The 10,500-degree solar surface emits mostly shortwave radiation with wavelengths near 0.5 micrometers. Earth, with its 59-degree average surface temperature, emits mostly 5- to 25-micrometer **longwave radiation**.

Searching for scientific answers

Soon after earning her PhD at the University of Washington and beginning her career at the prestigious National Center for Atmospheric Research (NCAR) in 1972, Peggy LeMone was selected to help lead aircraft operations during GATE, one of the biggest atmospheric research projects ever conducted. (GATE is a "nested" acronym for the GARP Tropical Atlantic Experiment; GARP was the Global Atmospheric Research Program.) GATE included twelve research aircraft operating out of Dakar, Senegal, collecting data over the Atlantic Ocean off Africa.

Unlike the gender-based discrimination that Joanne Simpson encountered early in her career, LeMone says she didn't encounter significant, overt bias during GATE or later during her NCAR career. When LeMone met Simpson in the early 1970s, Simpson "was delighted to meet another woman in atmospheric sciences," and was surprised by the number of women meteorologists LeMone knew. This led LeMone and Simpson to survey the 247 women meteorologists they identified and reported their findings in the February 1974 issue of the *Bulletin of the American Meteorological Society*. They concluded that conditions and opportunities for women "are improving"…but "difficulties must be overcome and compromises made if marriage, children, and an outstanding career are combined."

Peggy LeMone: After thirty years as a scientist, she's still finding new questions to ask about the atmosphere.

LeMone, who now holds the NCAR job title "Senior Scientist," says that as some of the old difficulties encountered by women scientists have faded; young male scientists who want to have families encounter many of the same problems as female scientists. Part of the problem then and now are those who think that eighteen-hour work days are needed to produce good science. LeMone disagrees, "As a scientist your mind never turns off. I'm two people, the creative one and the manager; the creative part is never in the office. Every time I stopped to have a baby or to have surgery I came back with new ideas."

Only a creative scientist, who has had the most advanced tools at her disposal to study weather as far away as Africa and the Solomon Islands for more than a decade, would decide to look for some answers in an ordinary puddle.

LeMone was pondering why a computer model designed to incorporate data about conditions on the ground in a weather forecasting model stumbled when puddles dotted the ground. In May 2007, she decided to observe a puddle down the street from her Boulder, Colorado, home. For five hours she recorded temperatures and cloudiness each half hour and took photos with chalk lines marking the puddle's edges as it evaporated, with alternating colors for different times. "I took pictures from the top of the puddle so I could superpose a **grid** on the puddle to estimate the area. But I forgot to record the depth! And in the end, I decided I should have been more systematic about recording wind speed. So this was, in some sense, a 'dry run'."

Whether or not her puddle observations lead to improvements in weather forecasting models, they provided a good topic for LeMone's Chief Scientist's Blog on the GLOBE (Global Learning and Observations to Benefit the Environment) Web site. GLOBE aims to help students and teachers around the world learn how to investigate the environment. As the part-time chief scientist for the program, LeMone aims to convince students that "you can be a field scientist at home."

She also has another message, "The best part of science is being surprised."

grees Fahrenheit, which in turn would radiate heat back into space. After sunset, the rocks continue radiating away heat, but with no solar energy arriving, the bare rocks cool to approximately −280 degrees.

Earth has a much smaller range of temperature extremes because its radiation balance is much, much more complicated. The oceans and the atmosphere enable the earth to balance incoming and outgoing radiation to produce temperatures at which life—at least life as we know it—thrives. The oceans are slow to warm up and to cool off. In brief, to warm a given amount of water by 1 degree, much more heat is needed than to warm the same amount of almost anything else (e.g., land) by a degree. In addition, vertical movements mix warm and cool ocean water.

The role of the atmosphere in keeping the earth at a livable temperature is more complex, as discussed by James Rodger Fleming in his *Historical Perspectives on Climate Change*. Fleming describes how in an 1824 essay, the French mathematician John Baptiste Joseph Fourier (1768–1830) advanced the concept of Earth's energy balance. He argued that atmospheric gases could reduce the amount of energy Earth loses via what we now call **infrared radiation**. In 1859 John Tyndall (1820–1893) performed laboratory experiments showing that several gases, including water vapor and carbon dioxide, absorb infrared energy, which causes the gas molecules to vibrate faster that is, to become warm. These warmer molecules radiate infrared energy in all directions, including toward the surface.

Molecules of these **greenhouse gases** continue radiating; they don't reflect heat. If they reflected heat, temperatures would plunge soon after sunset because there would be less energy for greenhouse gases to reflect. Greenhouse gases make the earth's average temperature approximately 60 degrees Fahrenheit warmer than it would be without them, because they heat the earth. In other words, the earth needs the greenhouse effect. The **greenhouse effect** that most climate scientists are concerned about can be considered the "enhanced greenhouse effect," caused by greenhouse gases that humans are adding to the air.

Calculating greenhouse numbers. During the nineteenth century, scientists discovered that Earth's climate had been much colder in the past; in fact, it had varied between frigid **ice ages** and warmer periods that we now call **interglacials** (we're now in one of these periods). Confirmation that huge amounts of ice once covered large parts of Europe and North America prompted a scientific quest to answer the question: What causes significant changes in global climate? Today's concerns about global warming and the changes that it will cause grew out of the search for answers to that question.

In 1896, Swedish scientist Svante Arrhenius (1859–1927) published one of the first plausible theories about what causes ice ages to begin and end. His complex calculations showed that cutting the amount of carbon dioxide in the air by half of nineteenth century amounts could cool high-latitude air enough to allow glaciers and ice sheets to grow. Arrhenius also realized that increasing carbon dioxide could warm the earth.

The seasons

You can trace all of Earth's weather back to the fact that the sun heats its surface unevenly. Unless you live in the tropics, where temperatures change little over the course of a year, you are accus-

---

A better name

Greenhouse gases don't act like a greenhouse, which keeps inside air warm by preventing convection currents from carrying heat away. But the term "greenhouse effect" has been around since the nineteenth century and we're stuck with it.

The "Callendar effect" refers to the direct relationship between amount of carbon dioxide in the air and temperature. It is named for Guy Stewart Callendar (1898–1964), the British scientist who laid the foundation for our current understanding of these gases in the 1930s.

Inside greenhouse gases

The molecules of greenhouse gases are made of at least three atoms. For example, carbon dioxide has one carbon and two oxygen atoms, water vapor has one oxygen and two hydrogen atoms, ozone has three oxygen atoms, and methane has one carbon and four hydrogen atoms. Only quantum mechanics can fully explain how this works, but in simple terms, the three or more atoms of a molecule of any greenhouse gas are bound loosely enough for them to vibrate more as they absorb infrared energy. Their vibration, in turn, gives off infrared energy. Molecules with only two atoms, such as molecular oxygen and molecular nitrogen, which make up approximately 98 percent of the air's molecules, aren't as free to vibrate and therefore do not emit infrared energy when they are warmed.

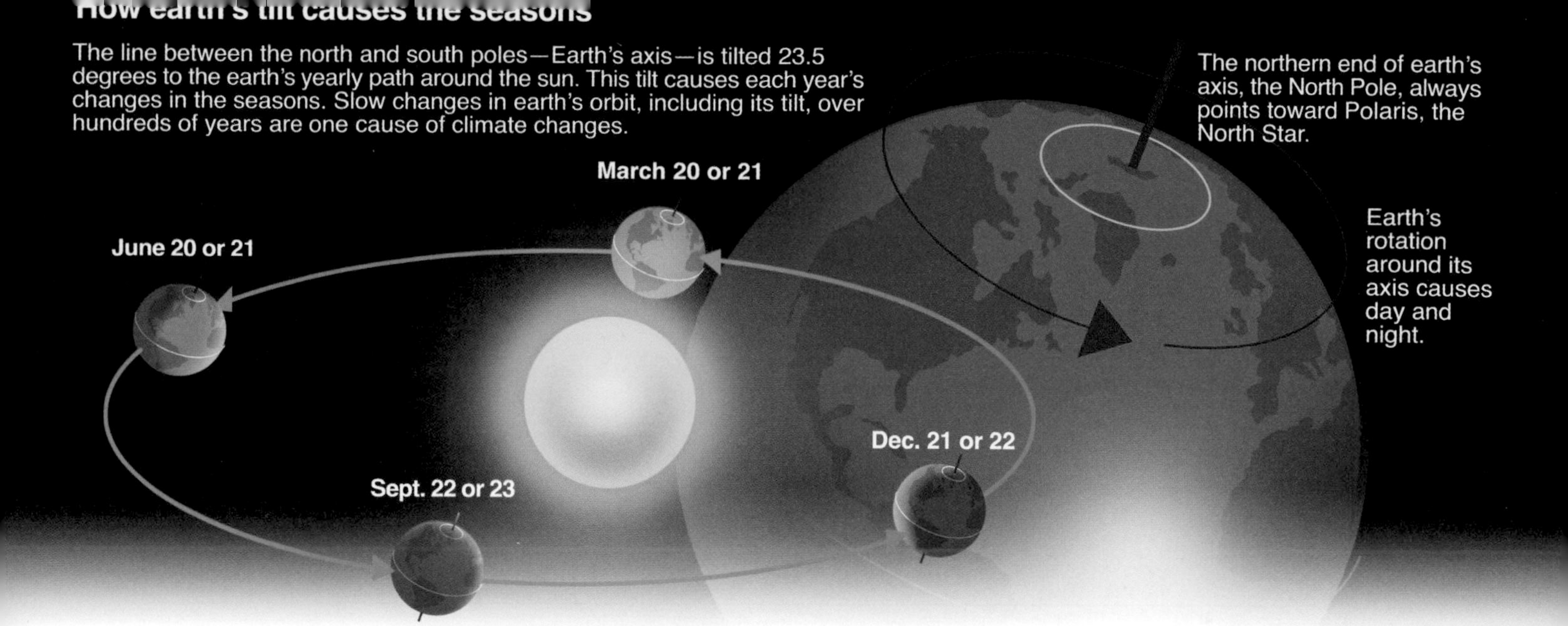

The line between the north and south poles—Earth's axis—is tilted 23.5 degrees to the earth's yearly path around the sun. This tilt causes each year's changes in the seasons. Slow changes in earth's orbit, including its tilt, over hundreds of years are one cause of climate changes.

The northern end of earth's axis, the North Pole, always points toward Polaris, the North Star.

March 20 or 21

June 20 or 21

Earth's rotation around its axis causes day and night.

Dec. 21 or 22

Sept. 22 or 23

tomed to temperatures varying according to season. Earth's axis—the line between the North and South poles about which the earth rotates—is tilted 23.5 degrees in relation to Earth's path around the sun. Winter days are short; the sun is lower in the sky than on long summer days. This effect is more pronounced with increasing latitude (closer to the poles).

Hawaii is the only U.S. state in which the sun is ever directly overhead, although in southern Florida it's less than 5 degrees from directly overhead the few days before and after the summer solstice.

U.S. National Weather Service seasons
Winter: December, January, February
Spring: March, April, May
Summer: June, July, August
Fall: September, October, November

To understand how this works, we'll consider what's going on in different parts of the earth on the solstices and the equinoxes. You often hear that the solstices and equinoxes are the "official" beginnings of the seasons. There's nothing "official" about these dates; they are the astronomical beginnings of the seasons. In fact, meteorologists in much of the world use three-month periods for seasons, as shown in the chart on this page. In tropical locations with monsoon climates, the real differences are between the wet and the dry seasons.

We'll start our seasonal journey around the world on the Northern Hemisphere's summer solstice.

On the summer solstice, no place north of the Arctic Circle is dark at any point during the twenty-four-hour period, while no place south of the Antarctic Circle will see the sun. In the Northern Hemisphere, the sun reaches its highest point of the year, while in the Southern Hemisphere it reaches its lowest point in the sky.

If you were at the North Pole for these twenty-four hours, you would see the sun make a complete circle around the sky, staying 23.5 degrees above the horizon all of the way around. Anywhere else north of the Arctic Circle you'd see the sun circle the sky following a tilted path. In Barrow, Alaska, or Tromso, Norway, for instance—both at 70 degrees north latitude—the sun will be 42.2 degrees above the southern horizon at solar noon (the time when the sun is highest in the sky) and 4.9 degrees above the northern horizon at solar midnight, when the sun is lowest in the sky.

South of Tromso, at 60 degrees north latitude, 448.5 miles south of the Arctic Circle, the sun sets briefly even on the summer solstice. The human population is much greater at this latitude, especially in Europe where it runs through Oslo, Norway. Places at this latitude have about 18 hours and 40 minutes between sunrise and sunset on June 21. On this day, all but an hour and twenty-four minutes of the five hours and twenty minutes between sunset and sunrise is **civil twilight**, and the rest is **nautical twilight**. Since the sky never really darkens, midsummer nights at such latitudes are known as "white nights," which are celebrated in the poetry and literature of St. Petersburg, Russia.

The terminator and twilight. Views of Earth from spacecraft show the boundary between light and darkness, the **terminator**, isn't a sharp line, because day *fades* into night, and night into day, during twilight. There are three kinds of twilight:

- Civil twilight: The sun is 6 degrees or less below the horizon. Most outdoor activities are possible unless it's cloudy.
- Nautical twilight: The sun is between 6 and 12 degrees below the horizon. You can make out the outlines of objects.
- Astronomical twilight: The sun is between 12 and 18 degrees below the horizon. The sun barely illuminates the sky overhead.

Less warmth from low sun. The time between sunrise and sunset is only part of the story of how the sun's path through the sky, caused by the earth's tilt, governs the amount of solar heating. The higher the sun is in the sky, the more it heats the earth. You can see part of the reason why by going into a dark room with a flashlight. Shine the flashlight directly down, and you'll see a circle of light, which has a certain amount of light energy hitting the floor. If you had enough information you could calculate the amount of light energy reaching each square inch of the floor. If you shine the light on the floor across the room, you'll see a stretched-out oval that covers more of the floor than the previous circle of light. Since the same amount of light energy is spread over a larger area, each square inch receives less light. The atmosphere also reduces the amount of solar energy reaching the earth when the sun is lower in the sky, because it goes through more of the atmosphere, which scatters away some solar energy.

At the latitude of Oslo, Norway, or Seward, Alaska, the sun climbs as high as 53.5 degrees above the southern horizon. The long days com-

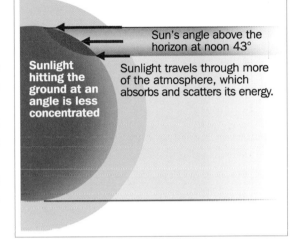

Longer days, less heat

On the summer solstice, June 20 or 21, the sun does not set on places north of the Arctic Circle, such as Barrow, Alaska, and Tromso, Norway, which are both at 70 degrees north. On the same day, the sun is up for only 13 hours, 40 minutes for places near the Tropic of Cancer, such as Key West, Florida, and Aswan, Egypt, which are both at 24 degrees north latitude. The images here show why Barrow and Tromso are much cooler, even in summer, than Key West and Aswan.

Barrow, Alaska, and Tromso, Norway

■ 70 degrees north latitude
■ No sunset

Sun's angle above the horizon at noon 43°

Sunlight hitting the ground at an angle is less concentrated

Sunlight travels through more of the atmosphere, which absorbs and scatters its energy.

Tables of sunrise and sunset times, such as those from the U.S. Naval Observatory, take the atmosphere's bending of light into account.

bined with the sun's higher angle in the sky supply much more solar energy to warm the ground and the water than during the winter when the sun is low in the sky. This is why summer is warmer than winter—more solar energy reaches the middle and polar latitudes of the summer hemisphere.

The difference in the height of the sun in the sky at noon and midnight for places north of the Arctic Circle, such as Tromso, Norway, explains why the times of day around noon are warmer than those around midnight—more solar energy reaches the ground around noon. Tromso, for instance, has an average high of 55 degrees Fahrenheit and an average low of 44 degrees on June 21.

The sun's height in the sky in the far northern and far southern latitudes explains why these regions aren't as warm as the tropics even though their summer days are much longer. For example, Oslo's average high temperature for the month of June is only 68 degrees, while Merida, on Mexico's Yucatan Peninsula in the tropics, averages 94 degrees in June, although the sun shines on it approximately six hours less each day than on Oslo.

Meteorologists re-calculate climate averages every ten years using the previous thirty years of records. The aim is to even out short-term swings while catching slower but steadier long-term shifts.

The earth's distance from the sun, which varies due to its elliptical orbit around the sun, plays only a minor role in temperatures. The earth is farthest from the sun in early July, approximately 94.8 million miles away. In early January it's closest, approximately 91.6 million miles away. But the earth's temperature, on average, is about 4 degrees Fahrenheit warmer in July than January, because land covers nearly 40 percent of the Northern Hemisphere and only 20 percent of the Southern Hemisphere. Since land warms up more quickly than water when both receive the same amount of heat, the Northern Hemisphere warms up more than enough to make up for being farther away from the sun during its summer.

Moving toward the equinox. After the summer solstice, Northern Hemisphere days grow shorter, while in the Southern Hemisphere days are growing longer until the September equinox. You often hear that the length of day and night are equal (12 hours each) all over the earth on the equinox, but that's not true. Day and night would be equally long on these days only if the earth didn't have an atmosphere. Earth's atmosphere refracts, or bends, light rays from the sun so that the top of the sun is still below the horizon when we see it rise. At the moment before the sun seems to completely set, its top is actually already below the horizon. This is why a day is roughly eight to ten minutes longer than twelve hours on the equinoxes. Depending on the latitude, day and night are of equal length for perhaps a week give or take a couple of days after the fall equinox and for a similar period of time before the spring equinox.

Warmth after the summer solstice. Even as Northern Hemisphere days begin growing shorter in late June, the weather continues growing warmer for another three or four weeks. In Minneapolis, Minnesota, for example, from 1971 through 2000 the year's highest average daily maximum temperature (84 degrees) occurred each day

---

Changes in light and dark during the year
The illustrations below and on the facing page show how Earth's pattern of sunlight and darkness change from season to season with the resulting changes in solar energy reaching all parts of the planet. The terminator—the line between day and night—moves across the warth from east to west (left to right in the images) at a steady pace of 15 degrees of longitude an hour.

Night and day on the summer solstice

The image shows where it's light and dark around the earth between 1 and 2 p.m. U.S. eastern daylight time on the summer solstice, which occurs on June 20 or 21.

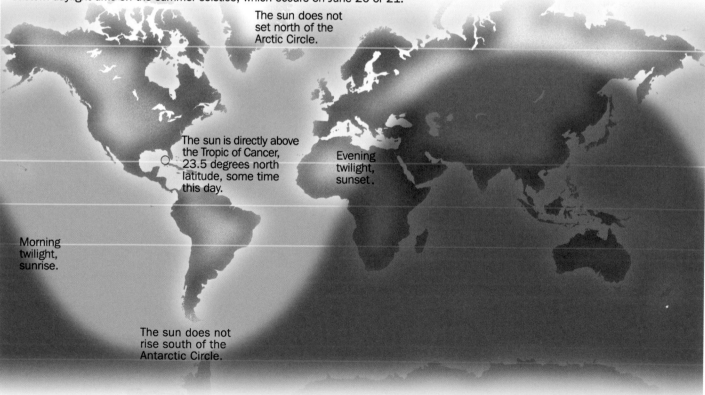

The sun does not set north of the Arctic Circle.

The sun is directly above the Tropic of Cancer, 23.5 degrees north latitude, some time this day.

Evening twilight, sunset.

Morning twilight, sunrise.

The sun does not rise south of the Antarctic Circle.

Night and day on the winter solstice

This image shows where it's day and night around the earth during the noon hour U.S. eastern standard time on the Northern Hemisphere's winter solstice, which occurs on Dec. 21 or 22.

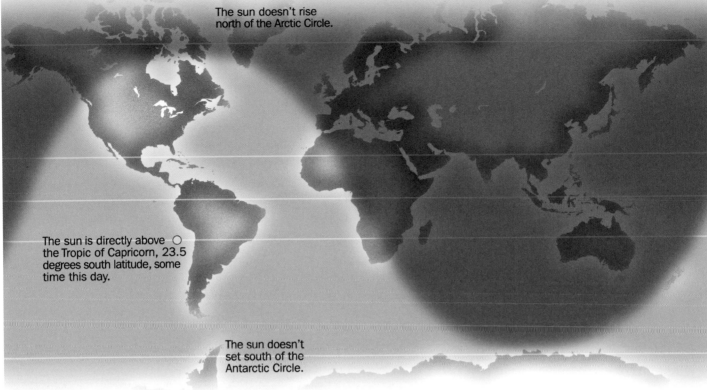

The sun doesn't rise north of the Arctic Circle.

The sun is directly above ○ the Tropic of Capricorn, 23.5 degrees south latitude, some time this day.

The sun doesn't set south of the Antarctic Circle.

Night and day on an equinox

This image shows where it's night and day around the earth during the noon hour U.S. eastern standard time on the Northern Hemisphere's spring equinox, which occurs on March 20 or 21, and between 1 and 2 p.m. eastern daylight time on the fall equinox, which occurs on Sept. 22 or 23.

March 20 or 21 and September 22 or 23 equinoxes

The sun rises at the North Pole a day or two before the March equinox and sets a day or two after the September equinox.

The sun is directly above the equator some time this day.

The sun sets at the South Pole a day or two after the March equinox and rises a day or two before the September equinox.

from July 15 through July 26, four to five weeks after the summer solstice. This lag exists throughout the summer hemisphere (and winter hemisphere) because the oceans, which cover 71 percent of the earth, heat (and cool) more slowly than land areas in response to seasonal changes in incoming solar radiation, and this has a direct bearing on heating and cooling of the atmosphere.

Of course, any particular day can be much cooler or warmer than average if cool or warm air arrives from elsewhere, which happens regularly in the middle latitudes, because masses of warm or cold air build up over various parts of the earth and then move elsewhere, bringing regular weather changes.

The coldest days of the year follow the year's shortest days by a few weeks (on average) as heat continues radiating away from the earth faster than heat is arriving, during short days with the sun low in the sky. In Minneapolis, the year's coldest average daily minimum temperatures, 3 degrees above zero Fahrenheit, occur between January 12 and January 16. Generally, a day's lowest temperatures will occur shortly after sunrise as the ground continues to radiate heat, but before the sun is high enough to make up the deficit.

Winter's unequal cold. As the earth continues in its orbit after the Northern Hemisphere's fall equinox, the sun's position in the sky continues moving south until the winter solstice.

Now, places at 60 degrees north latitude, such as Seward, Alaska, and Oslo, Norway, see the sun for only about five and a half hours a day. The sun climbs only 6.5 degrees above the southern horizon during the middle of the brief day on December 21. Both places are chilly but not as cold as you might think with the sun so low in the sky, and so briefly at that. In Seward, the average high and low December temperatures are 28 and 19 degrees Fahrenheit, respectively, while Oslo averages 32 and 22 degrees. Both places are near large bodies of water that take the edge off winter temperatures. Fort Smith, which is at the same latitude but far inland on the border between Canada's Northwest Territories and the Province of Alberta, shows how much difference this makes: its average December high and low are 1 degree above zero and 15 degrees below.

The solstices and equinoxes can occur a day before or after the dates given here, depending on the year and the time zone you're in.

Earth's climate regions. In talking about climate we describe the earth in terms of general climate zones, which are defined by the sun's path through the sky during the year.

The tropics are the places centered on the equator and extending north to 23.5 degrees north latitude—the Tropic of Cancer—and south to 23.5 degrees south latitude—the Tropic of Capricorn. The Tropic of Cancer is the northernmost point at which the sun is ever directly overhead, which happens on the Northern Hemisphere's summer solstice (June 21). The Tropic of Capricorn is the southernmost point at which the sun is ever directly overhead, which occurs on the winter solstice (December 22). The equinoxes occur on March 21 and September 23 when the sun is directly above the equator.

Unequal heating

If no other factors were involved, the tropics would continue growing warmer and warmer, because during the year this part of the earth receives more solar energy than it radiates away. The polar regions would grow colder and colder because even with the 24-hour days of summer, during the course of a year they radiate away more energy than they receive from the sun.

Simply stated, in actuality the tropics don't continue growing warmer and the poles colder because warm air rises in the tropics and cooler air moves from the polar regions across the surface to replace the rising air. If the earth did not rotate on its axis, the air that rises in the tropics would simply flow directly north and south all the way to the polar regions, where it would sink. It would then cool and flow toward the equator along the surface. In reality, however, the earth's circulation of air is much more complex with air movements and ocean currents balancing Earth's heat budget.

Climates don't exactly match regions. Most of the geographically defined tropics, except for the highest mountains, have tropical climates, which meteorologists describe as having continually high temperatures and considerable **precipitation**, at least during part of the year. Tropical climates extend north and south of the geographic tropics in many places to include areas such as southern Florida and the Bahamas.

The polar regions are the Arctic, which is centered on the North Pole and north of the Arctic Circle at 66.5 degrees north latitude, and the Antarctic, which is centered on the South Pole and south of the Antarctic Circle, which is 66.5 degrees south latitude.

The Arctic Circle is the farthest south you can go in the Northern Hemisphere and have at least one day when the sun doesn't rise—the winter solstice —and one day when it doesn't set—the summer solstice. In the Southern Hemisphere, if you headed away from the South Pole—where all directions are north—the Antarctic Circle is the farthest north you could go and still have one day a year with a midnight sun and one day with no sunrise.

While polar climates are cold, the Arctic doesn't always have the Northern Hemisphere's coldest temperatures. The Northern Hemisphere's coldest official temperature is −90 degrees Fahrenheit recorded in both Verkhoyanski and Oimekon in Siberia. Verkhoyanski is seventy-four miles north of the Arctic Circle, but Oimekon is 209 miles south of the Circle. The U.S. National Climatic Data Center's listing of the coldest temperatures officially recorded on each continent shows that the coldest Southern Hemisphere temperature, except in Antarctica, was the −27 degrees Fahrenheit recorded in Sarmiento, Argentina, at an elevation of 878 feet above sea level on June 1, 1907. Almost all of the Southern Hemisphere's land, except Antarctica, lies between 50 degrees south latitude and the equator. The ocean that surrounds Antarctica, with its moderating effect on temperatures, warms cold air moving toward Australia, New Zealand, South America, and Africa. In the Northern Hemisphere, large areas of land are between 50 degrees north latitude and the North Pole. The 50-degree latitude line runs just north of the U.S.-Canadian border, and in Europe it runs south of all of the United Kingdom, about three-quarters of Germany, all of Poland, and most of Russia.

Antarctica more than makes up for the lack of cold land elsewhere in the Southern Hemisphere. The world's record cold temperature of −127 degrees Fahrenheit was recorded on July 21, 1983, at the Russian Vostok Station in Antarctica (which means that Russians have recorded the official low temperatures for both hemispheres).

The middle latitudes, with their many varieties of weather and climate, are the places between the tropics and polar regions and have historically been called the **temperate zones** and as having **temperate climates**.

A little help from below. While solar energy does almost all of the work of driving the daily weather, **geothermal** heat from the earth's interior also contributes a small amount of energy that affects weather and climate. Going down into the earth from the surface, the temperature increases. For example, the *Future of Geothermal Energy Study* by an interdisciplinary panel led by the Massachusetts Institute of Technology that was released in January 2007 said that temperatures four miles under parts of the western United States reach 400 degrees Fahrenheit. This heat came from Earth's formation more than four billion years ago as iron meteorites, stony meteorites, and icy comets smashed together instantly converting the kinetic energy of their speed into heat. The pressure caused by the force of gravity consolidating the pieces into our planet added more heat and the decay of radioactive materials continues to add heat

Earth's subsurface heat announces itself most violently with volcanic eruptions. As with several other areas of science, Benjamin Franklin was one of the first to write about the possibility that volcanoes could affect weather. He speculated that the 1783 eruption of the Laki Volcano in Iceland could have caused "dry fogs" and unusually cold weather in North America and Europe that year. Today's scientists say he was correct.

Late twentieth century global weather data leave no doubt that the June 1991 eruptions of Mount Pinatubo in the Philippines sent 15 million tons of sulfur dioxide high into the atmosphere, creating a haze of tiny drops that blocked enough sunlight to cool the entire Earth by about 1 degree Fahrenheit for several months, as noted by the **Intergovernmental Panel on Climate Change**, which said in its 2001 report, "Mitigation of the [ongoing global] warming trend in the early 1990s was short-lived and was mainly due to the cooling influence of the eruption of Mount Pinatubo in 1991."

Past climates

Obviously, it is important to understand how climate changed in the past if we want to understand

Earth's energy budget

Solar radiation and its interactions with Earth's surface and atmosphere create a complex energy budget involving absorption, reflection, and radiation of various electromagnetic wavelength—predominately visual and infrared—that make Earth hospitable for life as we know it.

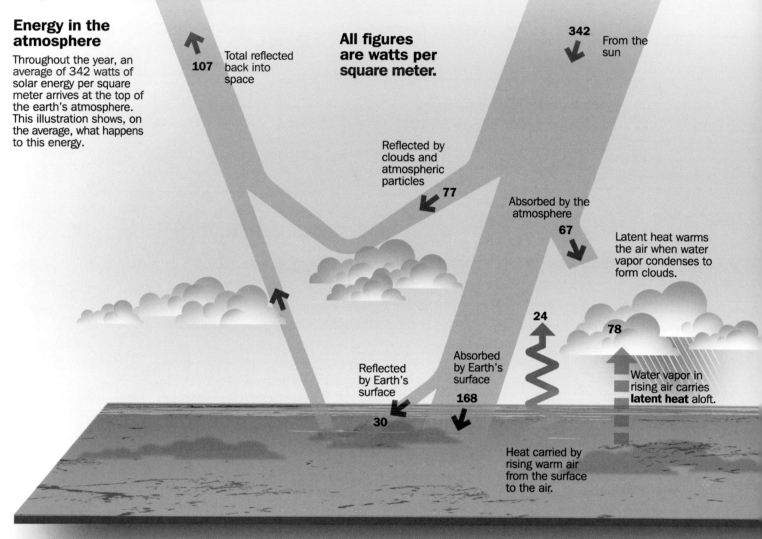

Energy in the atmosphere

Throughout the year, an average of 342 watts of solar energy per square meter arrives at the top of the earth's atmosphere. This illustration shows, on the average, what happens to this energy.

All figures are watts per square meter.

342 From the sun

107 Total reflected back into space

Reflected by clouds and atmospheric particles **77**

Absorbed by the atmosphere **67**

Latent heat warms the air when water vapor condenses to form clouds.

Reflected by Earth's surface **30**

Absorbed by Earth's surface **168**

24 Heat carried by rising warm air from the surface to the air.

78 Water vapor in rising air carries **latent heat** aloft.

Tropical heating, polar cooling

The illustration above shows the earth's average global budget of incoming and outgoing radiative energy. Since the sun is more directly above the tropics than the poles, the tropics receive more energy than they lose to space during a year. The map on the right shows these differences. Red colors indicate where more energy arrives than leaves. Yellow areas are close to being in balance. Green and blue areas lose more energy to space than they receive during a year.

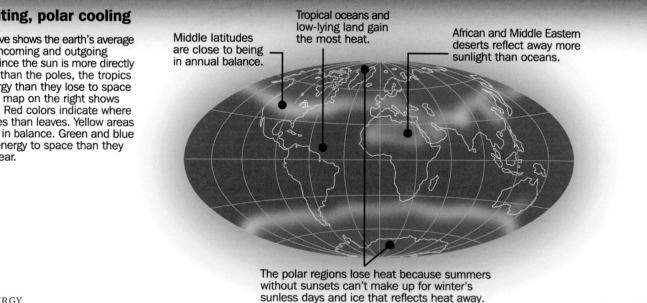

Middle latitudes are close to being in annual balance.

Tropical oceans and low-lying land gain the most heat.

African and Middle Eastern deserts reflect away more sunlight than oceans.

The polar regions lose heat because summers without sunsets can't make up for winter's sunless days and ice that reflects heat away.

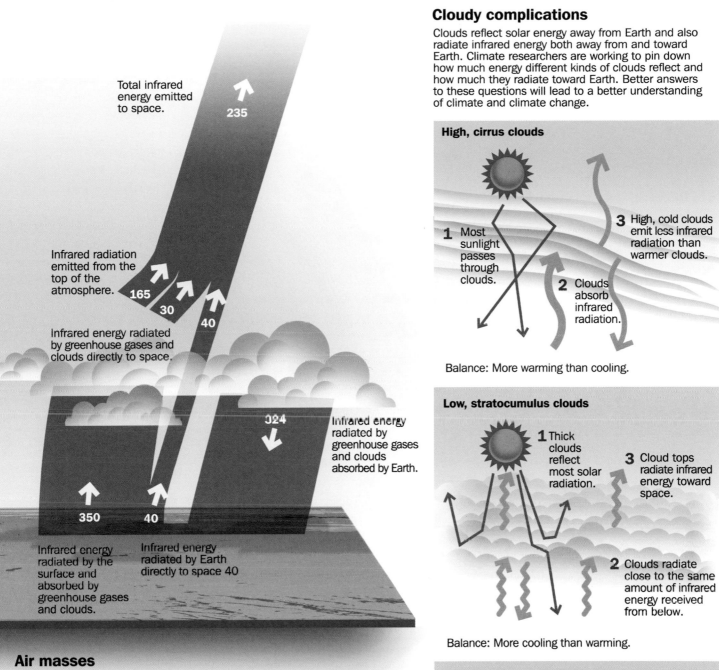

Total infrared energy emitted to space.
235

Infrared radiation emitted from the top of the atmosphere.
165
30
40

Infrared energy radiated by greenhouse gases and clouds directly to space.

324

Infrared energy radiated by greenhouse gases and clouds absorbed by Earth.

350
40

Infrared energy radiated by the surface and absorbed by greenhouse gases and clouds.

Infrared energy radiated by Earth directly to space 40

Cloudy complications

Clouds reflect solar energy away from Earth and also radiate infrared energy both away from and toward Earth. Climate researchers are working to pin down how much energy different kinds of clouds reflect and how much they radiate toward Earth. Better answers to these questions will lead to a better understanding of climate and climate change.

High, cirrus clouds

1 Most sunlight passes through clouds.

3 High, cold clouds emit less infrared radiation than warmer clouds.

2 Clouds absorb infrared radiation.

Balance: More warming than cooling.

Low, stratocumulus clouds

1 Thick clouds reflect most solar radiation.

3 Cloud tops radiate infrared energy toward space.

2 Clouds radiate close to the same amount of infrared energy received from below.

Balance: More cooling than warming.

Cumulonimbus clouds

2 Cold cloud tops radiate little infrared energy to space.

1 Thick, towering clouds reflect most sunlight.

3 Low cloud bottoms radiate close to the same amount of radiation received from below.

Balance: Close to equal warming and cooling.

Air masses

Large air masses form when air stays in a region for a few days; long enough to take on the characteristics of the surface it's sitting on. Continental air masses form over land or ice-covered oceans and maritime air masses form over ice-free oceans. Eventually, cold air masses move toward the tropics and warm air masses move toward the polar regions.

Air masses that affect North America

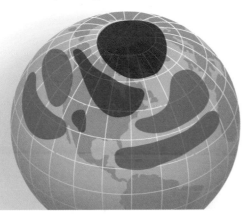

- Continental Arctic, dry, very cold
- Continental Polar, dry, cold
- Maritime Polar, damp, cool
- Maritime Tropical, humid, warm
- Continental Tropical, dry, hot

what could happen in the coming years and centuries. However, good records of day-to-day weather go back only to the middle of the nineteenth century, and the records don't cover some parts of the globe. Even without reports of weather measurements, we can discover much about weather and climate in past centuries by studying letters, diaries, histories, and government and business records. Such information takes us only so far; however, we need other ways to reconstruct climates of times before humans kept written records. In fact, we need to look back to times before humans populated the earth.

Information about prehistoric climates, in fact even some information about climate in recent centuries, comes from the work of **paleoclimatologists**. These scientists use data from **proxy climate sources**, such as cylinders of ice pulled from glaciers and ice sheets, or sediment from the bottoms of the oceans and lakes. ("Proxy" is sometimes used to refer to historical documents other than weather observations.) In Chapter 6, we describe the basics of the physics and chemistry behind some of the many methods for unlocking the secrets of past climates. Like all science, the methods and conclusions of paleoclimatologists have to withstand the knowledgeable criticism of fellow scientists. Arguments are often fierce, and established ideas are sometimes overturned.

A different Earth. During Earth's 4.5 billion-year history there have been long periods, sometimes lasting millions of years, when it has been much warmer than now, and similar stretches when Earth has been much colder. For example, some scientists advance the "Snowball Earth hypothesis," which argues that between 850 and 550 million years ago, Earth experienced the ultimate in ice ages with glaciers and sea ice even in the tropics. Other researchers agree that the earth had some very cold centuries during this period, but not *that* cold.

The evidence is much stronger that the Jurassic period, 206 to 144 million years ago, which saw dinosaurs living among thick strands of ferns and trees somewhat like today's palm trees, was much warmer than now. During past warm periods, sea levels and the amounts of greenhouse gases in the atmosphere were higher.

For the last million years, Earth's climate has alternated between ice ages, when temperatures and sea levels were much lower than now and ice covered more of the earth, and interglacial periods—we're in one now—with higher sea levels, warmer temperatures and less ice. The last ice age ended approximately 10,000 years ago, and since then Earth has enjoyed a period of relatively stable climate, though with some ups and downs.

These fluctuations have been significant enough to be at last partly responsible for the downfalls of some societies, such as the complex Mayan culture of what's now part of Mexico and Central America. The Mayan empire declined in the eighth and ninth centuries, perhaps in part because of prolonged droughts. Similarly, the Norse settlements in southern Greenland, which were populated from late in the tenth century into the fourteenth century, disappeared during the so-called **Little Ice Age**, a period of general cooling of Europe and other parts of the Northern Hemisphere that began in the middle of the twelfth century and ended in the middle of the nineteenth century.

Moving continents

The most amazing things early twentieth-century explorers found in Antarctica were coal (the remains of long-dead plants) and, in some of the few rocks that poke though the continent's ice, fossils of warm-weather plants. The only plants growing in Antarctica today are few and small and found on the Antarctic Peninsula (ice covers approximately 97 percent of the continent). The coal and fossils are evidence that earth's coldest continent was once very warm, either because the entire Earth was much warmer than any time in recorded human history, or because Antarctica had not always been at the South Pole—maybe both.

It turns out to be both. As we saw above, the earth has gone through periods when it was warm enough for forests and even dinosaurs to thrive in Antarctica. For instance, during the Cretaceous period, from approximately 144 million to 65 millions years ago, Antarctica's climate was somewhat like that of the U.S. Pacific Northwest today.

Antarctica was where it is now during the Cretaceous period, but it didn't arrive there until approximately 160 million years ago. This is one of the details of Earth's distant past that **geoscientists**, including climatologists, have worked out since the 1960s when they had found enough data of various kinds to convince them that the **plate tectonics** hypothesis was a good description of how the earth works, which is why it's now called plate tectonics theory.

The basic idea is that the earth's surface, including the ocean bottoms, is made of approximately a dozen large plates and more than thirty small plates that rest on a layer below, which over very long time periods acts like a liquid. The geothermal heat below the surface, which causes volcanoes and hot springs, supplies the energy that moves these plates at a speed roughly equal to the rate your fingernails grow, an inch or so a year. It's slow, but during Earth's long history these movements have made significant changes in the landscape and have been a major player in shaping the earth's climate.

Geoscientists have good evidence that approximately 250 million years ago, all of today's continents were shoved together into a supercontinent, called Pangaea, which stretched from pole to pole. Fifty million years later, Pangaea had separated into northern and southern parts—Laurasia in the north and Gondwanaland in the south. Laurasia eventually separated into Eurasia (Europe and Asia) and North America. Gondwanaland broke into today's continents of South America, Africa, Australia, and Antarctica. By the end of the Cretaceous period 65 million years ago, Antarctica was still attached to South America even though it was generally in its current location at the South Pole.

After Antarctica broke away from South America 30 to 40 million years ago, it became surrounded by ocean—the southern parts of the Atlantic, Pacific, and Indian oceans (often called the Southern Ocean). The steady west-to-east winds blowing over this unbroken ocean, which stretches around the world, created the Antarctic Circumpolar Current. Along with winds that circle Antarctica, it diverted much of the relatively warm water and air that ocean currents and wind transported toward Antarctica. As the globe cooled after the Cretaceous period, the ice sheet that now covers Antarctica began growing, killing the plants and animals there, leaving only the fossils that the explor-

Earth's moving continents
250 million years ago

All of Earth's land forms the super continent Pangaea.

65 million years ago

North America and Greenland
Eurasia
Africa
India
South America
Australia
Antarctica

As they drift apart, the continents begin to assume the shapes we know today.

When British explorer Robert Scott and his four companions died on the way back from the South Pole in 1912, they had 35 pounds of geological specimens on their sled, including a *Glossopteris* fossil. The specimens were found with Scott's body.

ers of the early twentieth century began finding (scientists are still finding such fossils).

One of the Antarctic fossils found by the early explorers was of the plant *Glossopteris*, which has been extinct for more than 200 million years. It has been found on every continent—a clue that all of the continents, at least those in the Southern Hemisphere, had once been together. (Its seeds would not have survived drifting across the oceans from continent to continent as those of some plants do.)

Arctic and Antarctic differences. Antarctica is by far Earth's coldest region—it's much colder than the Arctic. A little more solar energy reaches Antarctica than the Arctic because the earth is closer to the sun in January, when the sun never sets there, than in July, when it never sets on the North Pole. But this is not enough to warm it more than the Arctic.

Ice—more than two miles thick in places—covers approximately 97 percent of the Antarctic

An aurora australis above the U.S. South Pole Station on August 20, 2006. The sign and flag on the left mark the geographic South Pole. The first hint of sunrise is seen on the right, but the sun doesn't rise for another month. Auroras, including the aurora borealis in the Northern Hemisphere, are the most visible manifestation of the solar wind illustrated in the graphic on page 43.

large and neither are the changes since the 1645 to 1715 period that Eddy examined. In her *Physics Today* article Lean says the actual reduction in solar output then compared with today isn't known but "perhaps was as small as 0.05 percent."

Obviously, changes in solar output do not directly affect the Earth's temperatures in the same way that turning up a burner affects a pan of pasta you're cooking. Satellite observations, computer modeling, and other research are offering clues to how the sun-earth system leverages small energy inputs into large changes. For instance, as the number of sunspots increases, ultraviolet and x-ray radiation levels increase much more than visible light. Since ultraviolet energy affects mostly ozone and other gases high in the atmosphere, the sun's added energy heats the **stratosphere** and levels above it. Solar heating and cooling of the upper atmosphere affect atmospheric pressures with resulting changes in upper level winds.

Scientists are finding links between the highest parts of the atmosphere and some global

weather patterns. This brings us back to the Little Ice Age. In her *Physics Today* article, Lean describes research that suggests "a solar-induced stratospheric influence on the naturally occurring pressure oscillation in the Arctic and North Atlantic regions" at least helped cause the cold period.

Could increased solar output be causing global warming? By 2007 most scientists who study and sun and climate didn't think so. In its 2007 report, the Intergovernmental Panel on Climate Change (IPCC) says the warming of the past half century came during a time when the combination of changes in solar output and cooling from volcanic eruptions "would be likely to have produced cooling, not warming."

The sun's violence

While the sun's effects on weather and climate are subtle, the sun is anything but subtle when its corona rips open to send a coronal mass ejection (CME) with more than 200 billion pounds of electrons, protons, and atomic nuclei toward the earth at nearly the speed of light, dragging part of the sun's magnetic field with it. During periods when the sun is active it regularly shoots CMEs and accompanying solar flares—powerful bursts of high energy (X-ray) **photons**—toward the earth. Until the middle of the nineteenth century, however, no one knew about, or even suspected such events were occurring because the earth's magnetic field and atmosphere largely protects the planet from the sun's streams of deadly particles and high-energy photons.

Solar eruptions can cause havoc on Earth. The flare photons produce excessive **ionization** in the upper atmosphere that can cause communication blackouts. The CMEs create magnetic storms by pushing against Earth's magnetic field, causing the field lines to move. Whenever magnetic field lines move across an electrical **conductor**, they induce a current in the conductor. This is how magnetic storms create unwanted currents in power lines, sometimes knocking out power grids.

Today, the effects of CMEs and solar flares are known as space weather, and the joint Air Force–NOAA Space Environment Center in Boulder, Colorado, forecasts their effects as part of the NWS's National Centers for Environmental Prediction, which produces forecasts for what most of us think of as "weather," including hurricanes and tornadoes. The change recognized the growing importance of space weather and the growing ability to forecast it using new knowledge and technology.

Summary and looking ahead

The work of Tim Stanton and his colleagues at the North Pole described in this chapter's opening and of Judith Lean and her colleagues who probe how the earth and sun work together are just two examples of the many ways that scientists over the years have gone about trying to understand the flow of energy from the sun to and then around the earth. Understanding the flow of energy is basic to understanding any science. In the preface to her book *The Energy of Nature* (University of Chicago Press, 2001), E. C. Pielou nicely sums up the importance of energy: "Without energy, nothing would ever happen. Energy is as indispensable an ingredient of the universe as matter is." In a sense, the rest of this book is a more detailed look at the flow of energy through the atmosphere and oceans and how this affects all of the earth. In Chapter 3 we will see how the flow of energy affect the atmosphere and oceans, and in Chapter 4 we'll see how changes in water among its vapor, liquid, and ice phases is an important source of energy for the weather.

The forces that move the air and ocean water help to balance the earth's heat budget.

CHAPTER 3

As Hurricane Charley's winds began to roar through his canal-side neighborhood in Punta Gorda, Florida, on August 13, 2004, Jim Minardi took shelter in the home of his neighbors Don and Connie Farquharson. He says they "settled in with a really good bottle of California wine," and opened the shutters on the back door a couple of inches to watch the storm.

The storm sounded like a jet airliner's engines running full blast, Minardi says. "It was a banshee scream. That was just awesome to me. I was jaw agape, listening. It got so loud I was almost deaf afterwards. The gusts came with a 'wham.' It was just like a truck…we wondered 'what hit?' It was just a gust." Minardi was caught without shutters protecting his windows because he had spent the previous two days helping elderly neighbors install their shutters and lending a hand to other sailors trying to secure their boats. Like many others in Charlotte County on Florida's Gulf Coast, Minardi expected the hurricane to pass by out at sea on the way to hit Tampa to the north. By the time Charley was obviously heading for Punta Gorda with its winds strengthening to 140 mph, Minardi didn't have time to install his own shutters or flee.

As Charley's winds continued picking up speed, Minardi and his neighbors saw "trusses, roofing material, everything at once" from his house smash into Minardi's sailboat on the canal. "That was a sight. I can still see my boat and the mast going in the water, still intact when the roof material hit."

Charley left Minardi and his wife Teresa Fogolini, who was in California when the storm hit, with a house without a roof or windows. "What amazed me was that my refrigerator, which had been in the kitchen, was out in the backyard. I couldn't believe it would just pick up a refrigerator that was next to a concrete wall, move it around, unplug it, and send it out. It must have been swirling in there. All the walls were covered with leaf debris up to the ceiling." The house contained soggy clothing, mattresses, pictures, and papers; furniture was starting to rot; there was rust on every metal object; and mold was beginning to grow. "We tried to salvage some clothes, but there was no place to hang them up," Minardi says. "The computer was full of water, and when I turned it on it just smoked." The 40-year-old house couldn't be repaired; they would have to demolish it to build a new one.

A few days after the storm, Leslie Chapman-Henderson, the chief executive officer of the Federal Alliance for Safe Homes (FLASH), a nonprofit or-

Previous pages: This photo of Greensburg, Kansas, illustrating the destructive power of wind, was taken twelve days after one of the most powerful tornadoes on record destroyed 95 percent of the town's buildings on May 16, 2007.

ganization that promotes disaster safety and property loss mitigation, drove by and saw the wrecked house. She also saw the undamaged house of Chris and Jeri Webb across the street. It had been built in 2004 to exceed the more stringent building codes enacted after Hurricane Andrew hit Florida in 1992. FLASH used photos of the houses in a "Two Houses" publicity campaign describing why Charley hardly damaged one house while destroying one across the street. The Two Houses campaign turned out to be the best thing that could have happened to Jim and Teresa, as we'll see later in this chapter.

The air's pressure

Differences in air pressure cause the wind to blow, whether it's Hurricane Charley's 140 mph blasts or a light summer breeze. In this chapter we'll learn about air pressure and see how pressure differences and other factors, including Earth's rotation, create winds. Then we'll look at different kinds of winds, including **jet streams**. We'll also examine how winds cause ocean waves and help drive ocean currents. Ocean currents, like atmospheric winds, are an important factor in determining Earth's weather and climate.

Molecules are always moving, except at a temperature of absolute zero. The difference between the phases of matter—solid, liquid, or gas—is how strongly cohesive forces restrict molecular movement. When cohesive forces are strong enough to hold molecules in a definite shape, the substance is solid. A liquid's molecules are still bound together but freer to move, allowing it to take the shape of the container holding it. Molecules of a gas, on the other hand, are free from the cohesive forces that hold solids and liquids together and interact only when they collide.

At temperatures near the ground, air molecules are moving, on average, approximately 1,000 mph. If you put a lid on a jar without doing anything to the jar (such as heating it), you'd have a jar of air at the same temperature and pressure as the air on the outside of the jar. Inside, air molecules are zipping around, colliding with each other and the jar's sides, top, and bottom. We say the gas "fills" the jar because the moving molecules reach every part of it. But if you could somehow capture an image of the billions upon billions of air molecules in the jar at a particular instant, you'd see that the jar is mostly empty. The extremely tiny air molecules take up approximately one-tenth of one percent of the space inside the jar. That is, the air around you right now is about 99.9 percent empty space.

With so much empty space between the molecules of a gas, natural forces can easily compress and expand air, while liquids and solids, which don't have nearly as much room between their molecules, are extremely hard to compress. If air didn't compress and expand so easily, areas of high and low air pressure wouldn't form, and without them wind wouldn't blow.

Even though air is mostly empty space, fast-moving air molecules create pressure acting in all directions as they bounce off everything they hit. How much pressure they create depends on the number and mass of the molecules and how fast they are moving. At sea level, this pressure happens to be, on average, 14.7 pounds per square inch. This means that if you had a one-inch cube at sea level on an average day, the air would be pressing with 14.7 pounds of force on the top, the bottom, and the four sides of the box.

No matter how fast molecules are moving, weight—the force gravity creates—is always pulling them toward the center of the earth. Meanwhile, the rapidly moving air molecules are pushing in all directions, including up against their weight. The greater the weight pressing down on a gas, the more it's compressed. At the earth's surface, all of the weight of the air above compresses the air. The higher you go, the less compressed the air becomes. As the number of molecules in each cubic inch of air decreases with height, the pressure decreases. Air pressure approximately 18,000 feet above sea level is half that at the surface. In other words, half of the atmosphere's mass is below 18,000 feet.

Air density. Understanding the air's density is necessary in order to make sense of how the at-

> ### Tiny molecules
> Molecules are measured in nanometers. It takes a billion nanometers to add up to a yard. The air's molecules are one-third of a nanometer in diameter. At sea level, every cubic centimeter of air has approximately 100,000,000,000,000,000,000 molecules, yet 99.9 percent of that cubic centimeter is empty. A cubic centimeter is about the volume of the tip of your little finger.

Dry air is a mixture of gases that's about 78 percent nitrogen and 21 percent oxygen, with other gases accounting for the remaining one percent.

mosphere works. A substance's density is a measure of how much mass of the substance a particular volume contains, such as the number of kilograms in a cubic meter. Since you don't feel the air unless wind is blowing, you might be surprised by how dense it is. At sea level the air's average density is 1.225 kilograms per cubic meter. In units commonly used in the United States, we would say that a cubic foot of air weighs 0.077 pounds.

The density of the air changes with changes in atmospheric pressure and also with changes in temperature. In the atmosphere, if pressure stays the same, the air becomes denser as the temperature drops.

Imagine you have a column of air in the atmosphere. If you cool the air, the column grows denser by shrinking, which means that at any particular altitude, say at 10,000 or 30,000 feet, there will be less air above that altitude when the air is cold than when it's warm. The whole column of air shrinks when it cools. High-altitude pressure changes caused by heating and cooling of the air and the subsequent density changes create the pressure differences that cause jet stream winds to blow.

The air's density also depends on the pressure. You often hear about cold air being denser than warm air. This can be confusing to someone who is just beginning to learn about weather. "If cold air is more dense than warm air," such a beginner might ask, "and the air aloft is usually colder than the air at the surface, why doesn't the cold air aloft sink?" The answer is that lower atmospheric pressure and the resulting lower density aloft more than make up for the air being colder than the air below. For example, the air's average temperature at sea level is 59 degrees Fahrenheit, the average pressure is 1013 millibars, and the average air density 1.2 kilograms per cubic meter. At approximately 16,000 feet, the average temperature is zero Fahrenheit, and the average pressure is 540 millibars. This combination makes the air's density only 0.74 kilograms per cubic meter. Whether air sinks or rises

Pressure measurements

In this book, we generally use the metric measurement of air pressure, millibars, because American meteorologists usually use it. Most of the rest of the world uses hectopascals instead of millibars, but the figures are the same.

Standard sea-level pressure in different units:

1013.15 millibars or hectopascals
29.92 inches of mercury
14.7 pounds per square inch

Strictly speaking, attractive forces between molecules have no effect in an ideal gas but do have effects in real gases. Assuming that gases are ideal gases works for the temperatures and pressures meteorologists are concerned with and simplifies the mathematics of meteorology.

depends on its density in relation to the surrounding air at a given height.

Water, unlike air, is virtually impossible to compress, which means water pressure increases at a steady rate. That is, if you go twice as deep, the water pressure will double. Since water weighs so much more than air, you don't have to go down far to encounter extremely high pressures, which divers and submarine designers have to take into account.

Water density doesn't change as much as air density, but it does change. We see how this happens and how water density differences help drive ocean currents at the end of this chapter.

Dangers and challenges of high altitudes. Designers of high-flying airplanes need to find ways to keep the air pressure inside high enough for people to survive. This is why the cabins of such aircraft are sealed containers with air pumps that circulate fresh air. The pumps keep inside air pressure equal to the atmospheric pressure found between approximately 6,000 and 8,000 feet above the surface, no matter how high the airplane flies. If the airplane is flying at 35,000 feet, where the air pressure is about 3.5 pounds per square inch, and the pressure inside is the 11 pounds per square inch you'd find at 7,500 feet, the total pressure pushing out on the inside of the airplane is 7.5 pounds per square inch.

Painful pressure changes. We generally don't notice the air's pressure because it's almost always pushing equally in all directions. When something causes unequal pressure forces in our bodies, discomfort and, occasionally, life-threatening conditions can result.

Ascending or descending in a fast elevator or airplane causes pressure changes that you feel as ear blockages. Spaces in your middle ear contain air that's usually at the same pressure as the air around you. As outside pressure rises or falls, air flows in and out of your middle ear through the narrow Eustachian tube at the back of your nose. Quick pressure changes don't allow time for this flow to equalize inside pressure with the outside. If you're descending, the higher pressure outside pushes in on the eardrum, creating pain. If you're going up, air trapped inside the ear pushes out, which is also painful. A head cold or allergies can

How air and water pressure change as you go up or down

Note that the scales for air and water pressure shown here are different. The column of air on the left represents almost seven times the distance of the column of water on the right. Air pressures are listed in both pounds per square inch and millibars (the standard meteorological measure of air pressure in the United States).

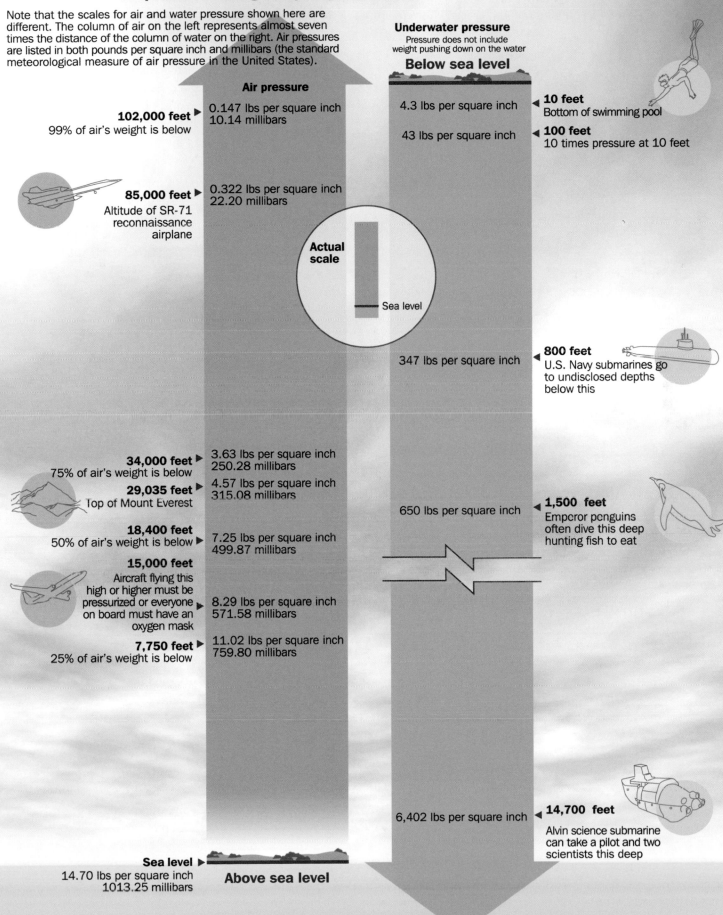

Air pressure

102,000 feet ▶
99% of air's weight is below

0.147 lbs per square inch
10.14 millibars

85,000 feet ▶
Altitude of SR-71
reconnaissance
airplane

0.322 lbs per square inch
22.20 millibars

Actual scale

— Sea level

34,000 feet ▶
75% of air's weight is below

3.63 lbs per square inch
250.28 millibars

29,035 feet ▶
Top of Mount Everest

4.57 lbs per square inch
315.08 millibars

18,400 feet
50% of air's weight is below ▶

7.25 lbs per square inch
499.87 millibars

15,000 feet
Aircraft flying this
high or higher must be
pressurized or everyone
on board must have an
oxygen mask

8.29 lbs per square inch
571.58 millibars

7,750 feet ▶
25% of air's weight is below

11.02 lbs per square inch
759.80 millibars

Sea level ▶
14.70 lbs per square inch
1013.25 millibars

Above sea level

Underwater pressure
Pressure does not include
weight pushing down on the water

Below sea level

4.3 lbs per square inch

◀ **10 feet**
Bottom of swimming pool

43 lbs per square inch

◀ **100 feet**
10 times pressure at 10 feet

347 lbs per square inch

◀ **800 feet**
U.S. Navy submarines go
to undisclosed depths
below this

650 lbs per square inch

◀ **1,500 feet**
Emperor penguins
often dive this deep
hunting fish to eat

6,402 lbs per square inch

◀ **14,700 feet**
Alvin science submarine
can take a pilot and two
scientists this deep

block the air passages to your sinuses. This is painful because air can't flow in or out to balance the pressure.

While quick changes in air pressure can cause pain, altitude sickness results when you stay too long in the low air pressure of a high elevation. How high you have to go to experience symptoms varies greatly among individuals, with some people becoming ill when they go higher than approximately 8,000 feet while a few people can get along at twice that elevation with no ill effects.

Oxygen continues to be approximately 21 percent of the air at higher altitudes, but the lower atmospheric pressure causes the problems. When you breathe in, the atmospheric pressure around you pushes air into your lungs. When the pressure is lower, the density of the air you take in is lower, which means each breath pulls in fewer molecules of oxygen than when the atmospheric pressure is higher. Every cell of your body, including those of the brain, needs oxygen to function; less oxygen begins to impair your body's functioning.

Most people can adapt to high altitudes, at least to places as high as Rocky Mountain ski areas. But this takes time, which means that if you live at a low elevation and visit a high-elevation ski resort, you should plan on taking it easy for the first day or so. Altitude sickness can be fatal. When at high elevations, if symptoms such as a headache, dizziness, or disturbed sleep don't go away after a day or so of taking it easy, or are joined by problems with balance or nausea, you should quickly get medical advice. Altitude affects people in different ways and physical fitness doesn't guarantee that you'll thrive at high elevations.

Measuring air pressure. Historians credit the Italian Evangelista Torricelli (1608–1648) with inventing the barometer in the early 1640s. Torricelli and others had made water barometers, but his breakthrough was to use mercury, a heavy metal that's liquid at room temperatures. The idea is the same for water or mercury. A tube filled with liquid is sealed at the top and sits open-ended in a container of the same liquid that's open to the air. You might think the liquid would run out of the tube into the container. It doesn't. Torricelli and others in the seventeenth century realized that air pressure pushes down on the liquid in the open container, pushing it up into the tube. If the liquid is water, the tube has to be at least 35 feet long to have empty space at the top. With mercury, which is approximately 13.5 times as dense as water, a 34-inch tube works, which means the instrument can sit on a table or hang from a wall.

Torricelli soon realized that a barometer does more than demonstrate that air has weight. The height of the mercury varies from day to day, sometimes from hour to hour. Temperature changes are one cause—the most common kind of thermometer uses mercury in sealed tubes—but changing temperatures don't account for all of the mercury's movements. (Weather observers today make corrections to barometer readings to account for temperature.) Observers soon realized that the mercury usually goes down before foul weather begins, while rising mercury normally signals fair weather. Figuring out the details of why this is the case took a couple of centuries.

In 1648, the French mathematician Blaise Pascal (1623–1662) came up with a hypothesis that logically follows from the idea that air has weight: The mercury should fall as you carry a barometer up a mountain, because as you go higher there is less air above you and thus less pressure pushing down on you.

Pascal and many others were developing modern science. One of the key differences between modern science and older ways of thinking is that with modern science, logical reasoning is not enough to prove theories. Rather, hypotheses must be confirmed through testing and experimentation.

Pascal wasn't healthy enough to hike up a mountain, but he persuaded his brother-in-law, Florin Perier, to carry a barometer to the top of the Puy-de-Dome, an extinct volcano just west of Clermont-Ferrand in south-central France, in September 1648. As Pascal had hypothesized, the mercury fell approximately 3.6 inches (in today's measurements) when Perier reached the 4,888-foot summit.

Why airplane doors won't open in flight
Opening the door of one of today's pressurized airliners when the airplane is in the air is impossible. Today's airliner doors must be pulled in at least a little to open because they are wider than the openings they cover.

The higher pressure inside the airplane holds the doors firmly shut when the cabin is pressurized. In a plane at 35,000 feet with 7.5 pounds per square inch of pressure pushing out, you'd have to pull with at least 3,705 pounds of force to open a 19-by-26-inch emergency exit during flight.

In this book we refer to "air" molecules instead of the cumbersome "molecules of nitrogen, oxygen, and other gases in the air."

In this book, we generally use American units for familiar quantities such as distances and speeds. But for quantities that are very small, such as the size of molecules, and quantities that you don't use in everyday life, such as the mass of the air, we use metric units.

How a mercury barometer works

3 Vacuum at top allows mercury to rise in sealed tube.

4 Observer lines up slider with top of mercury.

5 Observer reads height from scale.

2 Mercury rises in the glass tube.

1 Air pressure pushes down on mercury.

Mercury is shown in red to make it more visible.

Early scientists described air pressure as the height of the mercury in a tube, but by the late nineteenth century, meteorology was becoming a mathematical science. The great Norwegian meteorologist Vilhelm Bjerknes (1862–1951) persuaded scientists and weather services to begin using direct pressure measurements such as millibars (a metric unit similar to pounds per square inch) that scientists can easily use in equations. However, the International System of Units, adopted in the 1960s, uses the pascal, named for Blaise Pascal, for all pressure measurements, including air pressure. However, there is one problem: 101,315 pascals equals 1013.15 millibars. To avoid using such large numbers and to have figures equal to millibars, meteorologists decided to divide pascals by 100 and append the metric prefix "hecto." This makes the measurement, "hectopascals," abbreviated hPa.

In the United States, scientists generally use hectopascals or its abbreviation hPa in scientific papers, but the NWS reports surface air pressures in inches of mercury, the unit of measurement that's familiar to most Americans. However, the NWS uses millibars in its upper-air observations and forecasts, including upper-air charts. In general, in this book we use millibars for air pressures, but if you're used to the units used in most of the rest of the world, just substitute "hectopascals" for "millibars."

Other barometers

An aneroid barometer uses a small, sealed, flexible container with low air pressure inside. When the outside air pressure falls, the sides of the container bulge out. Rising pressure pushes them in. These movements operate a pointer that indicates pressure.

The electrical resistance in electronic barometer sensors depends on atmospheric pressure. Electronic circuits convert the sensors' resistance to digital pressure readings.

Pressure and winds

A little before 2 p.m. on August 13, 2004, an Air Force Reserve WC-130 hurricane hunter airplane confirmed the bad news from satellites and radar about Hurricane Charley: It was strengthening and heading toward Punta Gorda, Florida. Charley's estimated winds near the eye were 140 mph and surface air pressure in the eye was down to 954 millibars. Broadcasters reported this as 28.17 inches of mercury.

Even listeners who didn't understand air pressure and wind knew that the lower the barometric pressure was, the stronger the storm. Why? Air moves as wind from higher pressure toward lower pressure. The greater the pressure difference, the stronger the force pushing the air.

At 2 p.m. on August 13, 2004, the air pressure at Fort Myers, Florida (approximately 60 miles northeast of Charley's eye), was 1009 millibars, making the difference between there and the eye 55 millibars. The pressure difference between these two points was certainly not the only pressure force causing the winds. Higher pressures all around the storm were pushing air toward Charley's center. But to understand how wind works, it's much easier to look at how forces are acting on a single "parcel" of air. (Picture a balloon full of air or water; now take away the balloon but keep the air or water together. This is a "parcel" that you can follow as forces push it around and it changes.) In addition, showing

The seventeenth century English scientist Robert Boyle (1627–1691) coined the word "barometer" from the Greek words baros (weight) and metron (measure).

Fierce winds rip apart houses

Hurricane and tornado winds often destroy homes that seem to be well built, battering them not only with the force of the wind itself but also with flying debris.

1 The faster the wind, the more pressure it exerts on walls, windows, and doors.

2 Wind carries shingles, roof tiles, boards, and other debris that penetrate windows, doors, and even walls.

3 The wind creates an upward, lifting force on the roof.

4 Wind entering broken windows and doors adds to the upward push on the roof.

5 Flying debris can slice small trees; knock holes in houses, cars, and boats; and injure or kill anyone caught outside.

Charley destroyed the Minardi home. These debris missiles are the big danger, not only to other buildings, but also to people and animals anytime damaging winds blow.

So what causes walls of building to fall as they would if higher pressure inside a house had blown them down? Harper explains that you can think of a house's roof as a poorly designed airplane wing that creates a lifting force when a fast wind blows over it. This lifting force can pull a roof up from the walls. Most of the time, good connections between the roof and the walls and between the walls and the foundation should hold a building together. However, wind blasting through an opened window pushes up on the roof from the bottom, adding enough force to send it flying.

Harper says that all but two of the forty-two people the Wichita Falls tornado killed were in cars. "Thousands of people got into bathrooms, into bathtubs, and lived. The plumbing in the walls helps keep the walls from getting blown away." The Texas Tech researchers and architects, including Harper, "were pretty vocal on what we found. We started talking about it." Researchers started conducting wind tunnel tests.

As they examined the Wichita Falls damage, Harper and researchers from the Texas Tech Wind Science and Engineering Research Center found that construction quality often determined which houses the tornado destroyed and which it seemed to spare. Studies of tornado and hurricane damage continue to confirm this, leading to calls for stronger building codes, especially along the Gulf and Atlantic coasts where hurricanes bring strong winds to large areas.

The Wichita Falls findings, similar evidence from subsequent tornadoes and hurricanes, and wind tunnel tests at Texas Tech and other institutions are what led to today's National Weather Service advice that you shouldn't waste time opening windows. Instead, take shelter in a windowless room in a strong building. The Wichita Falls tornado also led to designs for "safe rooms" that can be built in houses, schools, or other buildings to protect inhabitants from flying debris.

You don't have to live along a coast that a hurricane could hit, or on the Great Plains of the United States in the area often called "Tornado Alley," to be concerned about tornadoes or other dangerous winds. After all, as Harper says, the tornado danger "isn't just in Texas and Oklahoma in the spring time." Tornadoes have hit each of the fifty U.S. states—although they are extremely rare in Alaska and Hawaii. In fact, tornadoes hit from time to time on every continent except Antarctica. A large share of the world's exceptionally violent tornados, however, hit the United States between the Rockies and Appalachians and in the Southeast.

No matter where you are, the basic wind safety rule is to take shelter in a sturdy building in a room without windows. There's always a chance that you'll have to react quickly, even in places where violent winds are rare.

Hurricane-resistant homes. As the fringe of Hurricane Wilma lashed Punta Gorda, Florida, on October 24, 2005, Jim Minardi and Teresa Fogolini relaxed in their new home, knowing it would protect them from the storm's wind and flying debris. Fourteen months before, Hurricane Charley had destroyed their old home. Now they were living in a house that was built as an example of hurricane-resistant construction.

The Federal Alliance for Safe Homes "Two Houses" publicity campaign featuring their destroyed home and the almost undamaged one across the street led Bob Vila, television's home-renovation guru, to do a series on building a hurricane-resistant house for Jim and Teresa. "I was weak in the knees; I couldn't believe it was happening," Mi-

Jim Minardi and his wife, Teresa Fogolini, on the sailboat that was hit by pieces of his house's roof during Hurricane Charley. Their new home, which was built on the site of the old one that Charley destroyed, is out of sight to the left. Behind them is the home of Don and Connie Farquharson where Minardi sheltered during Charley.

nardi says of the day in February 2005 he learned of Vila's decision.

In June 2005, with the new home's walls and roof completed, Minardi and William York, a consulting engineer, showed off some of its hurricane-resistant features. These begin with the reinforced concrete walls, which were poured in place, with the ends of the wall's reinforcing rods embedded in the concrete slab on which the house sits. The roof's rafters are firmly attached to the tops of the walls. To rip the roof away, the wind would have to also lift the walls and the slab. The house is a one-piece box built to withstand the battering of wind and flying debris. "I don't think this house is going to blow away," York says. "I think the walls would stand up to 200 mph winds."

The new house has a hip roof, instead of one with gable ends, which strong winds could demolish. The house's roof has five-eighths-inch plywood instead of the thinner half-inch plywood used in the construction of most houses. Even if wind doesn't rip away a roof, wind-driven torrents of rain can create leaks that turn a house into a mold farm. Thicker-than-usual layers of roof felt (tar paper) and tar go on top of the plywood to protect against leaks. Two stainless-steel screws, instead of the glue that's often used, hold down each of the roof tiles.

As Wilma approached, Minardi didn't have to worry about installing shutters on the new house, as its windows are impact-resistant glass enclosed

Gabled roof
If the gable isn't reinforced, wind can push it in.

Hip roof
Wind flows over the roof no matter what direction it comes from.

A One, Two, Three of Jet Streams

Pressure gradient forces and the Coriolis force determine the speed and direction of all winds. Friction also affects surface winds, but not winds aloft. On these pages we examine the forces that drive upper-atmospheric winds, including jet streams, which are often defined as winds aloft with speeds of 50 knots (57.5 mph) or faster.

1

Temperatures and pressures aloft

Pressures aloft are shown in millibars (mb) and temperatures in Fahrenheit degrees (F).

Southern weather station

Warm air expands, which means at any particular altitude, the pressure is higher than in a column of cold air.

Northern weather station

Cold air contracts, which means that at any particular altitude the pressure is lower than in a column of warm air.

2

The forces at 34,000 feet

PGF

Coriolis

The wind

West-to-east wind with forces in balance

Wind at 34,000 feet

The Coriolis force pushes the wind to the right.

The higher pressure 34,000 feet above the southern station creates a pressure gradient force pushing air toward the north.

3

Temperatures determine winds aloft patterns

Winter's large temperature contrasts create larger pressure differences aloft, and thus faster jet streams than summer's relatively small north-south contrasts.

Jet stream with frigid air over central U.S.

Coriolis force

Deep mass of cold air, low pressures aloft

PGF

Jet stream

Warm air, high pressure aloft.

Jet stream with frigid air over central U.S.

Coriolis force

Cool air, low pressures aloft

PGF

Jet stream

Warm air, high pressure aloft.

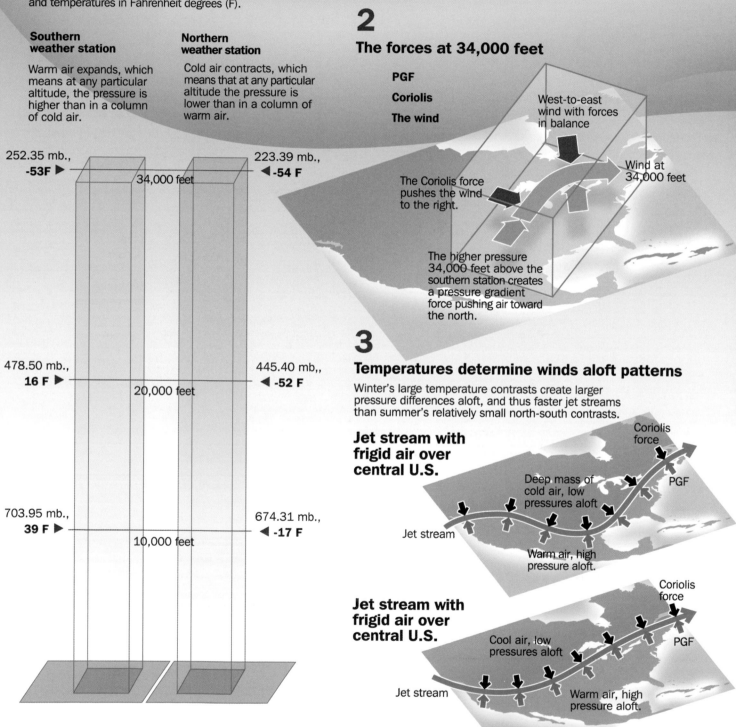

252.35 mb., **-53F** ▶

34,000 feet

223.39 mb., ◀ **-54 F**

478.50 mb., **16 F** ▶

20,000 feet

445.40 mb,, ◀ **-52 F**

703.95 mb., **39 F** ▶

10,000 feet

674.31 mb., ◀ **-17 F**

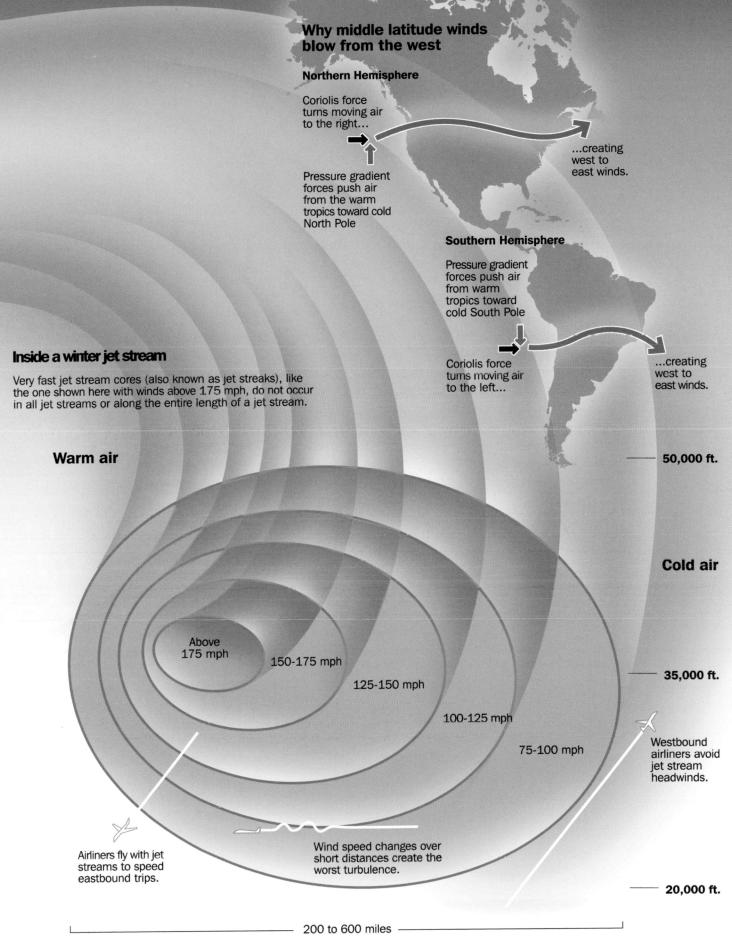

Why middle latitude winds blow from the west

Northern Hemisphere

Coriolis force turns moving air to the right...

Pressure gradient forces push air from the warm tropics toward cold North Pole

...creating west to east winds.

Southern Hemisphere

Pressure gradient forces push air from warm tropics toward cold South Pole

Coriolis force turns moving air to the left...

...creating west to east winds.

Inside a winter jet stream

Very fast jet stream cores (also known as jet streaks), like the one shown here with winds above 175 mph, do not occur in all jet streams or along the entire length of a jet stream.

Warm air

Cold air

— 50,000 ft.

Above 175 mph

150-175 mph

125-150 mph

100-125 mph

75-100 mph

— 35,000 ft.

Westbound airliners avoid jet stream headwinds.

Airliners fly with jet streams to speed eastbound trips.

Wind speed changes over short distances create the worst turbulence.

— 20,000 ft.

—— 200 to 600 miles ——

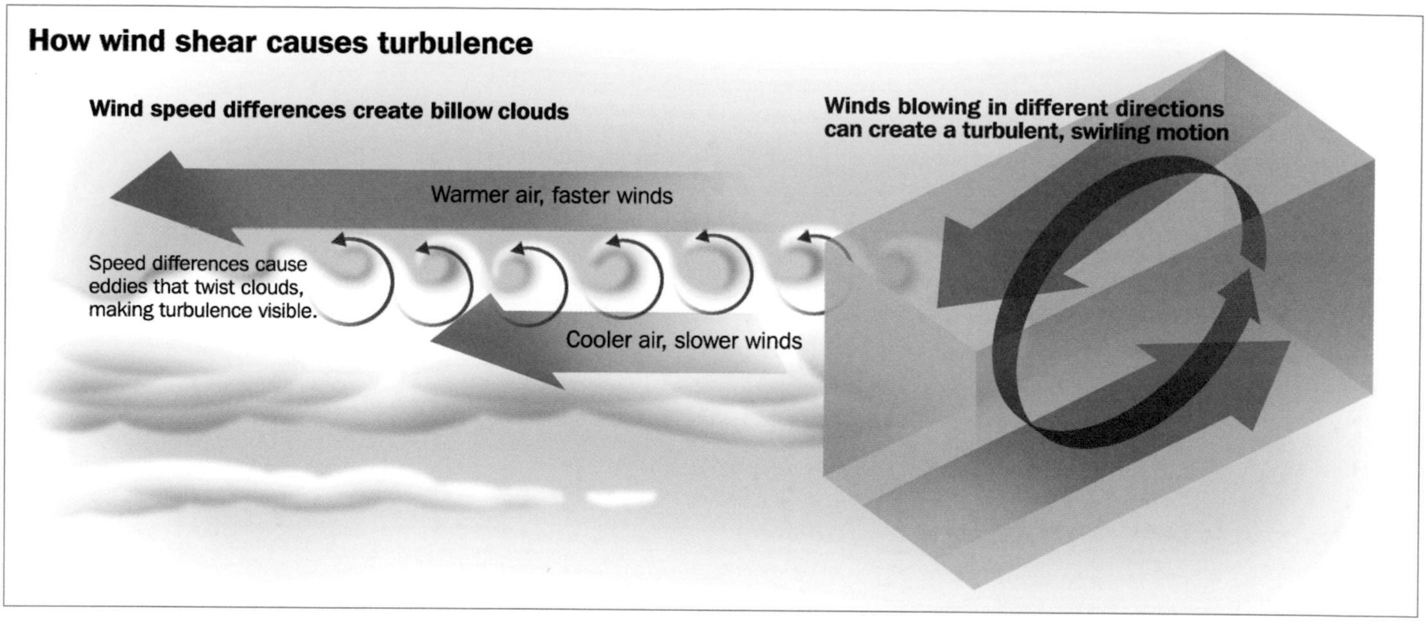

How wind shear causes turbulence

Wind speed differences create billow clouds

Warmer air, faster winds

Speed differences cause eddies that twist clouds, making turbulence visible.

Cooler air, slower winds

Winds blowing in different directions can create a turbulent, swirling motion

in frames designed to hold the glass in place when wind or debris hits.

The year before, Minardi had been amazed when he saw that the wind and flying debris hadn't broken the 15-foot picture window. Instead, "the frame and all came in, not just the window," he says. "I see how they do it now; I'm impressed. See these little screws," he says, holding a couple of one-and-a-half-inch wood screws like those from the old house's windows. "The old window had only three of these on each side." The new window frames are encased in the concrete walls, sealed against wind-driven water, and held by hardened, three-and-a-half-inch-long screws every 10 inches all around each window. Garage doors are another notorious weak spot, but not for the new house. Its garage door system, which has added bracing, was designed to stand up to winds of at least 150 mph, like the windows and other features.

York points out that even with the best design, "you are never one-hundred percent sure that there

will be no damage." A 140 mph hurricane could have gusts up to 175 mph or a small tornado. But even if the worst happens, hurricane winds won't destroy the new house and almost everything in it.

With this in mind, a standby generator will power the new house's air conditioning if a storm knocks out power for days, as hurricanes usually do. While air conditioning will help keep Jim and Teresa more comfortable in a storm's aftermath, York says its real value is defending against mold. "Ninety-degree, wet air is a great breeding ground for mold," he says. "One of the major problems we have is getting the electricity back on and getting air conditioning turned on to dry things out."

After bulldozers scraped away the last of the old house, dirt was trucked in to raise the lot to make the new home's floor nine-and-a-half feet above sea level. A three-foot-deep stem wall extends into the ground all around the slab. If a hurricane pushes storm surge in from the Gulf of Mexico, the stem wall should keep the water from undermining the house.

Minardi says he and his wife won't stay if a strong hurricane threatens to test the stem wall. One lesson he learned from Charley is, "Don't be so cocky; don't be so arrogant. We might have a storm-safe home, but where am I going to go if we have a 15-foot storm surge? My house will be [intact], although flooded to the roof top."

He and York thought about this for a few seconds before York said, "That's a lesson a lot of people don't learn."

Wind speed and pressure on structures

Wind pushes against structures, adding to the air pressure on the side the wind is hitting. As the wind speed doubles, the extra pressure increases by a factor of four.

Wind speed	Extra pressure (Pounds per square foot)
20 mph	1.60
40 mph	6.40
74 mph (Cat. 1 hurricane)	21.90
111 mph (Cat. 3 hurricane)	49.28
155 mph (Cat. 5 hurricane)	96.10
208 mph (F4 tornado)	173.06
261 mph (F5 tornado)	272.48

Bumps in the sky

On June 12, 1996, Air Force One was 33,000 feet above the Texas Panhandle with President Bill Clinton and about 70 others aboard when it suddenly bounced up and down several times, throwing the guacamole, tamales, frijoles, and salsa the crew were preparing for dinner around the cabin. No one was seriously injured, but according to Jill Dougherty of CNN, the cabin looked like a huge food fight had just ended. The incident is a reminder that the atmosphere doesn't care how important you are or how large an airplane you're in—Clinton's Air Force One was a Boeing 747—and that **turbulence** can catch even the most experienced and skilled pilots by surprise.

Airline pilots strive to avoid turbulence, but they are more concerned about passenger and flight attendant safety than aircraft damage. Today's airplanes are more than strong enough to stand up to any turbulence they're likely to encounter.

To form a picture of one cause of turbulence, especially at high altitudes away from the tops of huge thunderstorms and mountains, think of a whitewater stream with the water splashing up and down, creating waves and swirls. Since air isn't as heavy as water, any particular amount of force will create larger waves and swirls than the same force would in water.

Wind shear. Up and down air movements caused by rising warm air and sinking cool air cause most of the bumps that aircraft passengers and pilots feel at low altitudes. At the altitudes where jetliners cruise, almost all of the turbulence involves **wind shear**, which refers to winds that are blowing in different directions or different speeds, sometimes both in the same area. In later chapters, we'll see how small-scale wind shear close to the ground is a hazard to aircraft as they take off and land and how large-scale wind shear can weaken hurricanes. No matter the scale and location, wind shear gives the atmosphere a twisting motion that can cause eddies of swirling air.

Storm waves

December 21, 2005, was a great day to go to the beach in Southern California, especially if you were a skilled surfer. Temperatures climbed into the 70s,

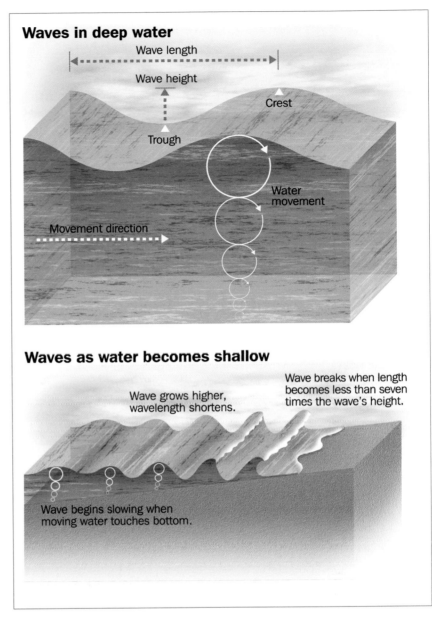

Waves in deep water

Wave length

Wave height

Crest

Trough

Water movement

Movement direction

Waves as water becomes shallow

Wave grows higher, wavelength shortens.

Wave breaks when length becomes less than seven times the wave's height.

Wave begins slowing when moving water touches bottom.

rain wasn't falling (December is during the region's rainy season) and waves in the 12- to 20-foot range were rolling ashore. The waves weren't to be trifled with; they washed a restroom off the end of the Venice Beach pier, prompting authorities to close other piers, parking lots, and sections of bike paths along the beaches. Larry Giles, a lifeguard at Encinitas, north of San Diego, told the *San Diego Union-Tribune* that during the day he saw "a lot of broken boards…a lot of broken egos."

Big ocean waves come from big storms, but as California surfers and others familiar with the oceans know, waves travel far and long after the storms that spawn them fade. The storm that sent surf crashing into Southern California in the days before Christmas 2005 churned up the ocean north

Water that waves push up a beach usually returns as a gentle backwash. At times, returning water becomes concentrated in rip currents that can carry swimmers into deep water. If a strong rip current catches you, don't swim against it; swim parallel to the shore until you are free of the rip current.

Chasing storms

Tim Marshall in Justin, Texas. The tower in the background houses the Doppler radar for station WFAA-TV in Dallas.

After he earned a bachelor's degree in meteorology from Northern Illinois University in 1978, Tim Marshall's dream came true: "The folks at Texas Tech University would actually pay me to study tornadoes," he recalls. "And I got to go storm chasing—I was like Indiana Jones, leaving the classroom to go hunt down the jewels."

Since 1978 Marshall has continued hunting the jewels, which, in this case are the answers to how tornadoes work and how wind destroys buildings.

His fascination goes back to the April 21, 1967, tornado that killed thirty-three people when it hit within a mile of his family's home in Oak Lawn, Illinois. Weather already had Marshall, then eleven, in its grip, and "I knew right then I wanted to study tornadoes."

After earning a master's degree in atmospheric science in 1980 and a master's in civil engineering in 1983, both from Texas Tech, Marshall started work at Haag Engineering in Dallas on August 1, 1983. Eighteen days later, Hurricane Alicia hit Galveston and Houston, showering downtown Houston with broken glass as flying debris shattered skyscraper windows.

"We learned that each building has a unique glass," he says. "We could identify the flying glass that broke windows of other buildings. We were able to pin down where it started." It's the kind of

forensic meteorology and engineering he's been doing ever since. "For every major hurricane, tornado, or hailstorm we can put together a wind and water map and learn what does the damage."

Haag gives Marshall four weeks off without pay each summer for tornado chasing. He's also been chasing hurricanes since August 1980 when, as a graduate student, he rode out Hurricane Allen in Corpus Christi, Texas. Video and photo sales pay for his storm chasing.

Storm chasers have been making scientific contributions since 1961 and will continue to do so. But as Marshall explained in *Weatherwise* magazine in 1993, more than the advancement of science is involved: "The most dedicated storm chasers... have a passion born of a need and love for the hunt. Chasing itself is a compelling test of strategy. The tornado is merely the bonus...There's no guarantee of success."

On May 9, 1982, Marshall and his wife, Kay, went tornado chasing on their first date and saw a tornado near Dawn, Texas, which Marshall photographed at sunset. Afterward she told him, "Hot damn, that was fun." Marshall says, "Right then and there I knew I was going to marry her."

While trying to catch tornadoes in action is his passion and hobby, Marshall's profession is learning how they and other storms destroy buildings.

For damage surveys, "I arrive right after these disasters," Marshall says. "I've never been able to get immune to each one of these disasters...I see the hardship...I can't get numb to it."

However, he says, "Tornadoes and hurricanes test buildings, and we can choose to learn from this or not. The damage left by Mother Nature is like a fingerprint. I like to analyze the failure to determine what failed first and why."

of Hawaii, more than 2,500 miles away, and had dissipated when its waves began reaching California. During the Northern Hemisphere's summer and fall, many of the big waves that make California a surfer's paradise come from fierce storms 7,000 miles away in the Southern Ocean around Antarctica, where it's winter, or from eastern Pacific hurricanes that never approach land.

Such waves do far more than entertain surfers or those who like to watch big breakers crash ashore on a sunny day. They transfer the motion of wind to the ocean, which helps drive the ocean currents that are a key part of the climate.

Huge, fast-moving waves. A tsunami can become a disaster when it meets the land. A tsunami travels at a speed related to the depth of the water. When a fast-moving tsunami from the open ocean approaches shore, water depth decreases and slows down the leading part of the wave. Because most of the tsunami is still traveling at high speed and since it hasn't yet reached the shore, the water can pile up to heights of several yards or more before coming ashore. Usually more than one wave washes in. They rarely arrive as typical breakers or a "wall of water," but are usually more like a sudden, extremely large rise in the tide, which is where the misnomer "tidal wave" comes from.

Sometimes, but not always, the water will retreat from the shore, exposing the sea bottom, in advance of the arrival of the crest of a tsunami. Knowing this has often saved the lives of people who flee inland to higher ground when the water moves away from shore. Feeling an earthquake when you are near the ocean should also be a sign to flee to higher ground.

On December 26, 2004, a violent earthquake along approximately 750 miles of a fault between two tectonic plates under the Indian Ocean sent waves estimated to be as high as 100 feet crashing into the coasts of South and Southeast Asia. Smaller but still deadly waves hit as far away as South Africa. The waves killed an estimated 275,000 people and made the word tsunami familiar to millions of people around the world.

Such earthquakes or large underwater landslides can create tsunamis that are 100 miles from crest to crest (in contrast to normal wind-driven waves with wavelengths more than 100 times smaller than that). While tsunami wavelengths are incredibly large, in the open ocean their height is minimal—less than a couple of feet high. Tsunamis race across the ocean faster than jet planes fly but can pass under ships without being noticed.

A basic wave primer. Ocean waves, including tsunamis, have important features in common with all other kinds of waves, such as sound waves, various waves in the atmosphere, and even the electromagnetic waves that we see as light and use for devices such as radios and radars. Waves transmit energy through a solid, a liquid, a gas, or even though a vacuum in the case of electromagnetic waves, producing regular back-and-forth, up-and-down, or circular movements of a medium's particles or fields of magnetic and electric forces.

An important point about waves: while the particles are moving back and forth, up and down, or in a circle, they don't travel with the wave. Only energy moves with the wave. For a simple example, tie one end of a piece of rope to something and hold the other end. If you move the end you're holding up and down or side to side, the motion travels the length of the rope as a wave. While each *part* of the rope moves up and down, the *entire* rope doesn't go anywhere. The wave in the rope is a **transverse wave** with the particles of the medium (the rope) moving at right angles to the wave's movement. Such waves move only through solids.

In a **longitudinal wave**, such as sound waves, the particles move back and forth in the direction of movement. Longitudinal waves can move through solids, liquids, or gases.

The particles of **orbital waves** move in circular paths. Orbital waves transmit energy only along the boundary between fluids—liquid or gas. Ocean waves are a good example. (Note that while in general use "fluid" usually refers to a liquid, scientists and engineers use it to refer to both liquids and gases.)

A closer look at ocean waves. All waves begin with a force that puts particles in motion and

Tsunamis and rogue waves
A tsunami could never turn over a ship at sea as in the 1972 disaster movie, *The Poseidon Adventure*. In the 2006 film *Poseidon*, they get it right, with a **rogue wave**.

Rogue waves have been blamed for many ship sinkings. In 1942, the British ocean liner *Queen Mary*, which had been converted to a troop ship, had more than 10,000 American troops and 1,000 crew members on board when a rogue wave, estimated at 92 feet high, hit it. It was 700 miles from the Scottish coast. The ship rolled 52 degrees, just short of enough to cause it to roll completely over.

A general definition of a rogue wave is an abnormally large wave compared to other waves at the same time and place. A few reliable, precise measurements have been made from ships and offshore oil well drilling towers. They have also been seen in satellite images.

Rich fishing areas
Marine biologists and other oceanographers continue to learn more about what the seventeenth-century whalers knew: sea life, including whales, is abundant along the edges of the Gulf Stream and other currents.

A current moving through the water sets up swirls and eddies along its edges, just as streams of air with different velocities stir up eddies along their boundaries. These areas are called **shear zones**.

Ocean swirls concentrate plankton, small fish, and invertebrates that attract larger fish, whales, and therefore the sailors that Franklin interviewed.

Focused on the oceans

The physics and poetry of the oceans enticed Lynne Talley into a profession that doesn't demand always working in an office.

The picture of global ocean currents and ocean effects on climate that we briefly describe in this chapter and in Chapter 5 comes from the work of oceanographers such as Lynne Talley, who applies the laws of physics to understand global currents and the formation of the ocean's water masses.

Today oceanography offers great career opportunities for men and women with a physics, applied mathematics, or engineering background who want to work out of the office on topics such as climate change. This is the case for Talley, who's a professor of oceanography at the Scripps Institution of Oceanography, part of the University of California San Diego, in La Jolla, California. Oceanography is much more open to women now than in the late 1970s when Talley became the fourth female graduate student to study physical oceanography at the Woods Hole Oceanographic Institution in Massachusetts, where she earned her PhD in 1982. Now, she says, about half of each year's top applicants for graduate studies at Scripps are women.

She cautions that young oceanographers, both women and men, who want to start families must cope with family and work demands. However, an advantage of working in academia, Talley says, is that "you don't have to punch the time clock, but I certainly put in the number of hours. You just do it evenings and can think about it pretty much anywhere, anytime." For instance, she can be at one of her son's soccer games, "thinking about which way the overturning circulation really goes."

Talley majored in physics at Oberlin College in Ohio and had internships in low-temperature physics, "which was very, very interesting." But when she heard about the Woods Hole graduate program she thought, "The ocean…that sounds really intriguing. You can write poetry about the ocean; it's a little harder to write poetry about liquid helium." A chance to travel was also an attraction, but she cautions that working on an oceanographic ship is not as luxurious and interesting as you might think. The ships are far from cruise liners, and they spend little time in exotic ports.

Some oceanographers don't go to sea, but Talley finds it important. When she's on a ship watching data as it comes in, she sometimes sees things that prompt her to look at a problem in a new way or point her research in a new direction. "Because you're at sea, you have fourteen hours a day to work. Everything is taken care of. Your laundry is right there; they feed you meals. You don't have to play cribbage and watch movies, you can work. I love it."

Talley graduated from Oberlin in 1976 with a bachelor of music degree in piano performance in addition to her physics degree. She became a scientist instead of a musician because "you can make a living at physics. If I had stayed with piano as an occupation, I would mainly be giving lessons, but what I really like to do is play."

In her "second job as a church pianist" for a Unitarian Fellowship, she gets to play for Thursday choir practice and every Sunday morning. She also plays dual piano chamber music with a friend, with whom she has entered amateur competitions. "Given that I enjoy doing what I do, I'm glad to do music on the side. I get an intellectual high from doing physics problems, but I get my emotional highs playing Mozart. It is necessary to me."

Atlantic sinks to become what is called North Atlantic Deep Water, which along with **Antarctic Bottom Water** is part of the global system of ocean currents. Scientists have used various names for this system of ocean currents, such as the "great oceanic conveyor" or "conveyor belt" because they carry nutrients and other matter from place to place, somewhat like a conveyor belt moves parts across a factory floor. The system, particularly the vertical component, is also called the **thermohaline circulation** from the Greek words for heat and salt. This refers to the fact that density differences caused by temperature and salinity changes drive the vertical circulations. By the beginning of the twenty-first century, many oceanographers and climatologists were using the term **meridional overturning circulation (MOC)** to describe this phenomenon. Meridional refers to the north-south flow of ocean water, and overturning refers to vertical current. Many prefer MOC because it recognizes that not only density differences but also winds and other forces drive the system.

A major driver of the meridional overturning circulation is in the northern Atlantic near Greenland. To see how this works, let's follow a cubic foot of water from the Gulf of Mexico to the northern Atlantic via the Gulf Stream. In the Gulf of Mexico, the water could be 80 degrees Fahrenheit and approximately 3.5 percent salt, close to the average salinity for the world's oceans. This water has a density of 63.854 pounds per square foot.

As the water travels north, water vapor evaporates from it, which means it becomes saltier (when water molecules leave the ocean as vapor, the salt stays behind). Rain falling on the Gulf Stream makes up for some of the evaporation but not all.

In addition to transporting water vapor, winds blowing over the ocean carry away some of its heat. If the water's salinity stayed at 3.5 percent, but it cooled to 40 degrees, its density would be 64.113 pounds per cubic foot. But, in addition to cooling, the water grows saltier from evaporation as it moves north. Assume it reaches the ocean off Greenland with 3.8 percent salt. This, combined with the 40-degree temperature, would make its density 64.316 pounds per cubic foot

These changes aren't much, but they are enough to cause the denser water to sink deep in the ocean and flow back toward the south as an underwater current. As water sinks in the northern Atlantic, more water flows north to replace it. If northern Atlantic water grows denser either from increased evaporation, less rainfall, or additional cooling, water will sink faster in the northern Atlantic and the Gulf Stream will speed up to replace the sinking water. Therefore, the sinking dense water acts like a motor driving the conveyor belt.

The sinking of the water can also slow down. Anything that makes the water less salty will slow the sinking. This can occur if more rain or snow fall on the Atlantic Ocean or if more Arctic Ocean sea ice or icebergs that break off from Greenland drift into the Atlantic and melt. Both could be consequences of global warming.

In addition, not all of the water that sinks in the northern Atlantic goes all of the way to the bottom. Some of it sinks a few hundred feet and flows into the Arctic Ocean as underwater currents that circle around below the surface of the Arctic Ocean. Water in these currents takes perhaps a quarter of a century to circle around the Arctic Ocean and merge back into the Atlantic. Changes in the temperature or salinity of this water could speed up or slow down the meridional overturning circulation. This possibility has led to speculation that a slowdown of the MOC could cool Europe, because the Gulf Stream would bring less warm water north to warm the winds blowing across Europe.

Summary and looking ahead

The destruction by Hurricane Charley's winds described in the opening of this chapter is a dramatic illustration of why people want to understand wind. The importance of winds, however, goes far beyond their power to destroy. The weather and ocean currents help the earth balance its heat budget. In this chapter, we learned a little about how they do this, beginning with our examination of the causes of winds and ocean currents and some of the things that winds and ocean currents have in common. We learned about some of the effects of wind and how to protect yourself and your home from strong winds.

Now we're ready to move on to Chapter 4 where we will examine the role of water in all of its forms in creating both weather and climate as winds and ocean currents move it around the globe.

It's ordinary but complex and needed for life—and without it our weather would be dull.

CHAPTER 4

The thought of being personally involved in the phenomena you study attracts many men and women to atmospheric or oceanic research. That, and the likelihood of discovering something that could help people, such as by improving forecasts, put F. Martin (Marty) Ralph in the chief scientist's seat behind the pilot in the cockpit of a WP-3 research airplane in the middle of a Pacific storm off California on February 2, 1998.

Scientists who study the atmosphere, hydrosphere, oceans, or the earth's ice depend on the vast amounts of data collected by observation tools, such as ocean buoys and weather instruments. Usually the observations from ocean currents, ice sheets, or storms are transmitted to scientists who can be thousands of miles away. At times, however, scientists need to be on the scene to make minute-by-minute decisions about what to measure. This is why oceanographers go to sea on research vessels, polar scientists trek across ice sheets, and meteorologists fly into storms, as Ralph and his colleagues were doing in 1998.

Images from satellites hundreds of miles above the the earth give scientists information about what's going on in storms over the oceans, but clouds hide important details. Until they approach very close to land, such storms are out of the range of ground-based weather radars. Using satellite images and other data plus the basic physical laws of meteorology, scientists developed hypotheses about what happens as Pacific storms come ashore in California. The only way to confirm such hypotheses, however, is to go into the storm to measure winds, air's vertical movements, air pressures, humidities, and temperatures, and determine how these change as the storm moves ashore. An airplane, especially the NOAA WP-3, with its three weather radars and variety of weather instruments, is an incredibly effective way to collect the needed observations.

As director of the California Land-Falling Jets Experiment, or CALJET, Ralph's airborne responsibilities included knowing what the scientists aboard the airplane were seeing on the radars and other instruments, keeping an eye on the weather outside, and, on any given flight, "looking at these in the context of the scientific hypothesis we were trying to test."

On February 22, 1998, they were flying into a Pacific storm to study how the geography of North America's west coast creates unique weather condi-

Previous pages: Rain from one of the West Coast storms NOAA scientists were studying in February 1998 triggered a mud slide that damaged homes and vehicles in Rio Nido, California. The resort community is on the Russian River in Sonoma County, north of San Francisco.

Flying weather labs

NOAA's two Lockheed WP-3s are the same kind of airplane the U.S. Navy has been using to patrol the world's oceans since 1962. NOAA's two WP-3s are best known for their hurricane research flights, which they have been doing since June 1976.

The airplanes are based on the design of the Lockheed L-188 Electra, which carried as many as 127 passengers for many airlines in the 1960s and 1970s. (Only a few small airlines use them today, mostly for cargo.)

NOAA's WP-3s have a crew of eight or nine and can accommodate a dozen others, including scientists and, on some flights, news reporters and photographers. During flights, which can last a little longer than eleven hours, everyone has a seat with at least as much room as in an airliner's first-class cabin. Most of the seats have computer monitors and keyboards for scientists to use during flights. Except when the pilots illuminate the big "Fasten Seat Belts" sign,

Flight meteorologist Jack Parrish (center) and Systems Crew Chief Sean McMillan (left) brief Commerce Secretary Donald Evans at a work station aboard a WP-3. NOAA is part of the Department of Commerce.

those aboard can wander into the cockpit to look over the pilots' shoulders or relax in the galley in the rear with a snack—bring your own—or freshly brewed coffee.

---

tions. A storm moving ashore from the Pacific first crosses the mountain ranges that parallel the coast, then a fifty- to eighty-mile-wide valley, and finally it traverses the high mountains of the Sierra Nevada or Cascade ranges. The resulting weather is much more complex than the weather from ocean storms that move across the generally flat, wide coastal areas of eastern North America or Western Europe.

"It was a very exciting time to be out there with this incredible platform, looking at how the atmosphere and ocean interact and how storms behave as they hit the coast," Ralph says. A bonus was that the view included "some of the most beautiful mountains. I feel so privileged to have had an opportunity to see the world that way, from a scientific viewpoint. If you saw something that you weren't expecting, you could actually go after it."

The CALJET scientists expected that the storm details they uncovered would lead to better forecasts of when and where precipitation will fall and how much is likely when a Pacific storm slams into California. On this particular flight, the scientists wanted to test the hypothesis that cold air blowing through mountain passes and out over the ocean could push up warm, humid air from the tropical

Pacific heading toward the coast. If the weather worked as they thought it did, clouds formed in the rising air would dump heavy rain on the cities, towns, and highways along California's coast. If the incoming humid air didn't begin rising until it hit the mountains a few miles inland, the heaviest rain would fall on less-populated areas.

"We were down to about 100 meters (approximately 300 feet) above the ocean," Ralph says, describing the flight that had gone out over the ocean and was heading back toward land. "Out the front window there's a storm hitting the coast. As we were heading toward the coast, there was what looked like a fog bank up ahead in a line. We punched through that fog bank and in a second the wind shifted" from blowing toward the coast to blowing parallel to the coast.

Ralph asked the pilots to fly low over the coast to collect measurements. "I'll never forget looking down on the pine trees right below us. It was pretty intense." The WP-3 then flew back out over the ocean and then returned to the coast, collecting data that confirmed the previous readings.

The wind blowing parallel to the coast was a "barrier jet," created by cold wind blowing through

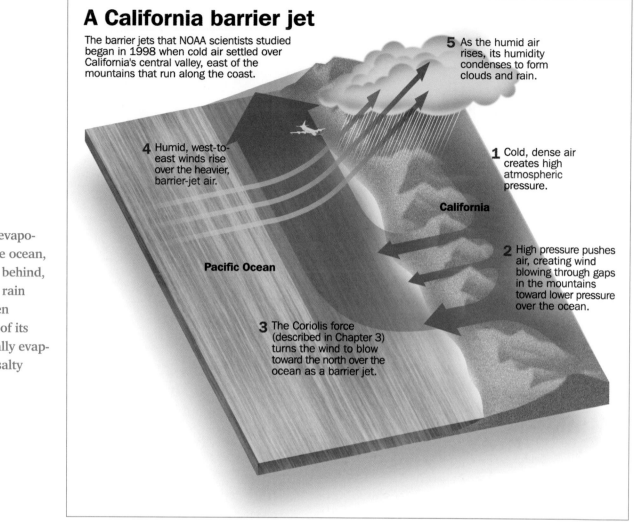

A California barrier jet

The barrier jets that NOAA scientists studied began in 1998 when cold air settled over California's central valley, east of the mountains that run along the coast.

5 As the humid air rises, its humidity condenses to form clouds and rain.

4 Humid, west-to-east winds rise over the heavier, barrier-jet air.

1 Cold, dense air creates high atmospheric pressure.

California

Pacific Ocean

2 High pressure pushes air, creating wind blowing through gaps in the mountains toward lower pressure over the ocean.

3 The Coriolis force (described in Chapter 3) turns the wind to blow toward the north over the ocean as a barrier jet.

When water evaporates from the ocean, the salt stays behind, which is why rain isn't salty even though most of its water originally evaporated from salty oceans.

the mountain passes. The Coriolis force turns such winds to the right, to blow toward the north in this case, parallel to the coast. When the warm, humid air encountered the barrier jet's cold, dense air, it rose. The water vapor in the rising air was condensing into clouds that spread rain along the coast.

After the second flight through the barrier jet confirmed it was pushing the humid air up, Ralph and others exchanged high fives as they stood in the aisle between the airplane's workstations. "We had just discovered something," Ralph says. "During my career there have been a handful of times when I can say I have found something new. I can say I found something that really is new" that day.

Water and weather

To understand what the CALJET researchers learned, you need to know a little about the many ways water helps create weather. When you look

at a weather forecast, you're usually most interested in **precipitation**. Is it going to rain on your picnic? Is snow going to clog your route to work? The importance of water to weather goes far beyond rain and snow, however. Invisible water vapor is always in the air. When there is a relatively large amount of water vapor in the air on a hot day, you notice it as humidity. When water vapor condenses it can create dew, fog, and clouds as well as rain and snow. As we will see in this chapter, water is even the fuel that powers thunderstorms and hurricanes. To make sense of The earth's weather, you need to understand the role of water.

Water's importance goes beyond weather. It's needed for all of the forms of life that we know about. Some only need a little, but all need water.

For something that's so common that most of us rarely think about it, water is really quite strange. If you happen to be sipping a glass of ice water as you read this, you are experiencing one of

How evaporation saturates air

The water and air in each container below warm or cool to the same temperature. The lid, which moves freely up and down, keeps air above the water from mixing with the outside air.

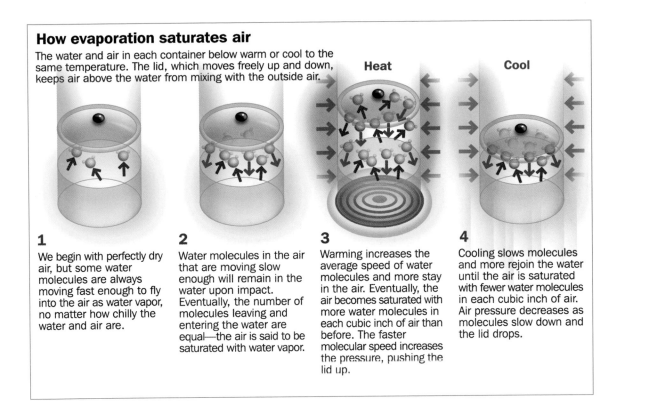

Heat

Cool

1
We begin with perfectly dry air, but some water molecules are always moving fast enough to fly into the air as water vapor, no matter how chilly the water and air are.

2
Water molecules in the air that are moving slow enough will remain in the water upon impact. Eventually, the number of molecules leaving and entering the water are equal—the air is said to be saturated with water vapor.

3
Warming increases the average speed of water molecules and more stay in the air. Eventually, the air becomes saturated with more water molecules in each cubic inch of air than before. The faster molecular speed increases the pressure, pushing the lid up.

4
Cooling slows molecules and more rejoin the water until the air is saturated with fewer water molecules in each cubic inch of air. Air pressure decreases as molecules slow down and the lid drops.

---

water's strangest aspects: It's the only natural substance that can exist as a liquid, a solid, and a gas at room temperature. The water in the glass is a liquid, ice is a solid, and the air you're breathing contains water vapor. To understand why water acts this way, we need to look at water molecules and the atoms of oxygen and hydrogen they are made of.

Even if you never studied chemistry, you probably know that water is called H_2O because each water molecule consists of two atoms of hydrogen (H) and one atom of oxygen (O). These atoms lock together as a molecule because the oxygen atom shares electrons with the two hydrogen atoms in **covalent bonds**. The sharing isn't equal, however. The oxygen atom attracts electrons with more force than the hydrogen atoms. Since electrons are negatively charged, this gives the oxygen side of a water molecule a small negative charge. The hydrogen atoms, which are both on the same side of the oxygen atom, have a weak

A water molecule

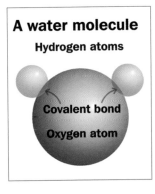

Hydrogen atoms

Covalent bond

Oxygen atom

positive charge. Thus, water molecules are "polar," with negative and positive sides.

Since opposite charges attract, water molecules tend to stick together with the hydrogen sides attracted to the oxygen sides of other molecules in what chemists call a **hydrogen bond**. These bonds are much weaker than covalent bonds and thus easier to break.

Molecules of all substances are always moving; the higher the temperature, the faster they move. When you heat water, some of the heat's energy goes into breaking hydrogen bonds as well as accelerating water molecules. This is why more heat is needed to increase water's temperature by a particular amount than to increase the temperature of most other liquids by the same amount. Water also takes longer to cool, because, as its temperature decreases, new hydrogen bonds begin forming, adding heat to the water and slowing cooling. Water's "reluctance" to warm up and cool down moderates the the earth's climate.

Water in the air. The fog bank the CALJET scientists saw off California's coast on February 2, 1998, formed when water vapor, which had evaporated from the ocean hundreds of miles away, condensed into tiny droplets of water. Since evaporation and condensation are so important to

Lab experiments using extremely clean air have created **supersaturated air** with relative humidity as high as 800 percent. In nature, supersaturation normally doesn't reach more than a fraction above 100 percent relative humidity because even relatively "clean" air contains plenty of condensation nuclei.

weather, we'll begin our exploration of water with the phase changes from liquid to vapor and back into liquid. Then we'll move on to the phase changes involving ice before discussing how these changes affect weather.

The illustration on the previous page will help you understand what happens when water evaporates and water vapor condenses. In the illustration, molecules of water and air in the containers are moving at a range of speeds clustered around an average speed. The amount of water vapor in the air when it becomes saturated depends on the temperature. You sometimes hear that "warm air can hold more water vapor than cold air," but it's not a matter of air "holding" water vapor. Scientists discovered in the late eighteenth century that water vapor is not dissolved in air in the way that salt is dissolved in ocean water.

Temp (°F)	Saturated mixing ratio (g/kg)
90	30.95
80	22.18
70	15.73
60	1.03
50	7.63
40	0.19

In relation to their size, the billions upon billions of air's molecules are far apart. For all practical purposes, they don't react with each other, which is another reason for not talking about air "holding" water vapor—molecules in a gas aren't holding anything, and they are so far apart that there's plenty of room for relatively few water molecules. Warm saturated air has more water vapor because as the temperature increases, more molecules gain the speed needed to break free of liquid and become vapor (at a higher energy state), while fewer vapor molecules are going slow enough to condense into liquid (at a lower energy state).

Meteorologists commonly measure the amount of water vapor in the air using grams of water vapor per kilogram of dry air. They call this the **mixing ratio**. The maximum amount of water vapor in the air for a particular temperature is the **saturated mixing ratio**. The table above shows saturated mixing ratios for a few selected temperatures. We'll use this table to help you understand ordinary humidity measurements.

The first thing the table shows is that the amount of water vapor needed to saturate the air increases rapidly as the temperature rises. For instance, the amount of water vapor needed to saturate the air increases by more than four times when the air temperature doubles from 40 to 80 degrees Fahrenheit.

Measuring humidity. The table to the left helps show what the common humidity measures of **dew point** and **relative humidity** mean. Let's say that at 2 p.m. a weather observer measures a 90-degree temperature and calculates, based on other measurements, that the air contains 7.63 grams per kilogram of water vapor. If the amount of water vapor in the air stays the same as the temperature cools to 50 degrees that evening, the air would be saturated. What happens when the air becomes saturated? Water vapor begins condensing if it has something to latch onto. For instance, water vapor condenses onto grass to form dew. This is why the temperature at which a particular mass of air will become saturated is known as the dew point. Above the ground, water vapor condenses on tiny particles known as **condensation nuclei** to form fog or cloud droplets.

In other words, at 3 p.m. the weather observer could have reported a temperature of 90 degrees and a dew point temperature of 50 degrees, based on the air's actual mixing ratio at the time. Dew point is nothing more than the temperature that a particular mass of air has to reach in order to be saturated. When air is sitting over land or cool water, the mixing ratio, and thus the dew point, generally change little unless a new mass of drier or more humid air arrives. If the arriving air has stayed over a warm ocean, such as the Gulf of Mexico for hours or even days, the dew point increases as water continues evaporating into the air. If moist Gulf of Mexico air is being pumped northward into the central United States, the dew point for the region increases as the warm, humid air arrives. If a mass of air stays over cold land long enough, it not only cools but also becomes drier as water vapor precipitates out as snow.

Even with the same air mass, the mixing ratio can vary during the day. Observations show that if no wind is blowing, the ground-level mixing ratio can be higher during the morning because water vapor that's evaporated collects near the surface. As the ground warms up during the day, air near the ground rises and drier air from above descends.

Weather services generally don't list official dew point extremes, but the U.S. Army Corps of Engineers' 1997 booklet *Weather and Climate Extremes* lists the world's highest average afternoon dew point as 84 degrees Fahrenheit on the Red Sea coast of Eritrea in northeast Africa. In his book, *Extreme Weather*, Christopher Burt reports that Dhahran, Saudi Arabia, recorded a dew point of 95 degrees Fahrenheit on July 8, 2003.

This can lower the mixing ratio. Late in the afternoon, as warm air stops rising, the mixing ratio rises again. These changes, however, are small compared to those caused by the arrival of new air masses.

Dew point is the best, rough measure of how uncomfortable humidity will make you feel. No matter what the temperature, air with a dew point of about 60 degrees Fahrenheit or higher will feel humid to most people. When the dew point hits 70 degrees, most people feel that the humidity is un-

Temp (°F)	Saturated mixing ratio (g/kg)	Actual mixing ratio (g/kg)	Relative humidity (percent)
90	30.95	7.63	25
80	22.18	7.63	34
70	15.73	7.63	48
60	11.03	7.63	69
50	7.63	7.63	100

comfortable, and a dew point above 75 degrees makes the humidity oppressive.

The relative humidity measures how much water vapor is actually in the air compared with the amount that would be in the air if it were saturated at the current temperature. To find the relative humidity, you can divide the air's actual mixing ratio by its saturated mixing ratio, and then multiply by 100 to make it a percentage. (This is not the exact scientific calculation, but it's close enough for most purposes.) To see how this works, we'll show how the relative humidity changes as 90-degree Fahrenheit air with a mixing ratio of 7.63 cools to 50 degrees with the actual mixing ratio staying the same. Remember, we said that at 3 p.m. the air's mixing ratio was 7.63 grams of water vapor per kilogram of air. As the air cools, this mixing ratio stays the same; no water is evaporating into the air or condensing out of the air. To calculate the relative humidity when the temperature is 90 degrees, you divide 7.63 by the saturated mixing ratio for 90-degree air (30.95) and multiply the result by 100 to obtain 24.65, which rounds off to a relative humidity of 25 percent.

The first thing the table shows is that as air cools and the amount of water vapor stays the same, the relative humidity increases. This shows why relative humidity, which many people refer to as "the humidity," is not as good a measure of how uncomfortable you'll be on a warm day as dew point is.

The table also shows that air with 100 percent relative humidity can have less water vapor than air with a lower relative humidity. For instance, with the air and dew point both at 50 degrees, the relative humidity is 100 percent with only 7.63 grams of water vapor per kilogram of air. But if the dew point were 60 degrees and the temperature were 90 degrees, the relative humidity would be 36 percent—sounds low, but the amount of water vapor in the air is 11.03 grams per kilogram, which is more than in air with 100 percent relative humidity at 50 degrees.

Often on a sticky day people will say things like, "the temperature and humidity were 95." Such heat-humidity combinations don't happen in the United States. For a 95-95 temperature-relative humidity combination, you'd need a dew point of 93 degrees, which you'd find only near shallow, very warm seas, such as the Red Sea and the Persian Gulf. During especially brutal heat waves, dew points in the United States climb into the 80s but not into the 90s.

When the dew point is 70 degrees, most people complain about the humidity. But a temperature of 90 with a dew point of 70 gives you a relative humidity of "only" 51 percent, which sounds low. Take another case. One February morning Nome, Alaska, had a relative humidity of 91 percent. You can be sure no one outdoors in Nome that morning complained about the humidity: The temperature was 9 degrees and the dew point was 7 degrees.

Ice and life

When 39-degree water sinks it carries nutrients and oxygen down with it, which support life during the winter. The survival of life in frozen lakes also depends on the fact that ice is a good insulator, which helps prevent the water below from cooling enough to freeze and kill plants and animals.

If very cold water and ice were heavier, they would sink, covering the bottoms of lakes, ponds, and oceans—including tropical oceans—with ice. Aquatic life would be cut off from plants and tiny animals on the bottoms of today's lakes and oceans. Probably even more important, if ice formed at the bottoms of deep lakes and the oceans, spring and summer sunlight wouldn't melt it. In such a case, we can be sure that life on Earth would be much different and may not have evolved into the many complex forms that now populate the the earth.

How upper-air temperatures make air stable

When a weather balloon finds that air cools relatively slowly with altitude, the air is stable. Temperatures are given in Fahrenheit.

What happens when stable air is pushed up

Air is forced up when it's in the center of an area of low air pressure at ground level, when wind pushes it over mountains or when warm air rides over colder air.

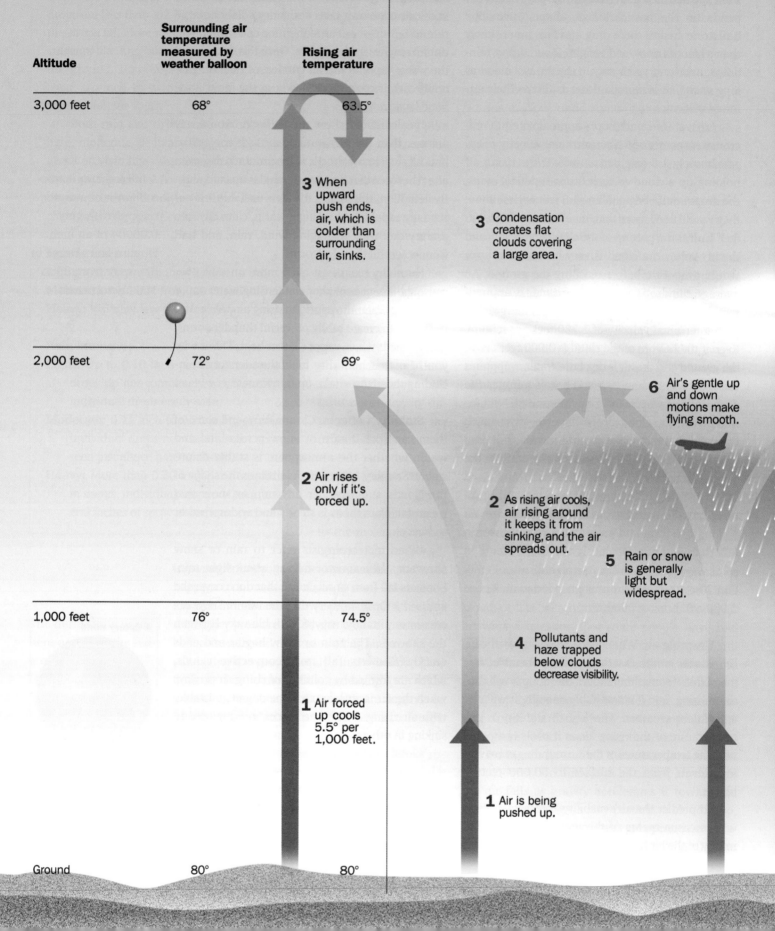

Altitude	Surrounding air temperature measured by weather balloon	Rising air temperature
3,000 feet	68°	63.5°
2,000 feet	72°	69°
1,000 feet	76°	74.5°
Ground	80°	80°

3 When upward push ends, air, which is colder than surrounding air, sinks.

2 Air rises only if it's forced up.

1 Air forced up cools 5.5° per 1,000 feet.

3 Condensation creates flat clouds covering a large area.

6 Air's gentle up and down motions make flying smooth.

2 As rising air cools, air rising around it keeps it from sinking, and the air spreads out.

5 Rain or snow is generally light but widespread.

4 Pollutants and haze trapped below clouds decrease visibility.

1 Air is being pushed up.

How upper-air temperatures make air unstable

When a weather balloon finds that air cools rapidly with altitude the air is unstable. Temperatures are given in Fahrenheit.

What happens when unstable air rises

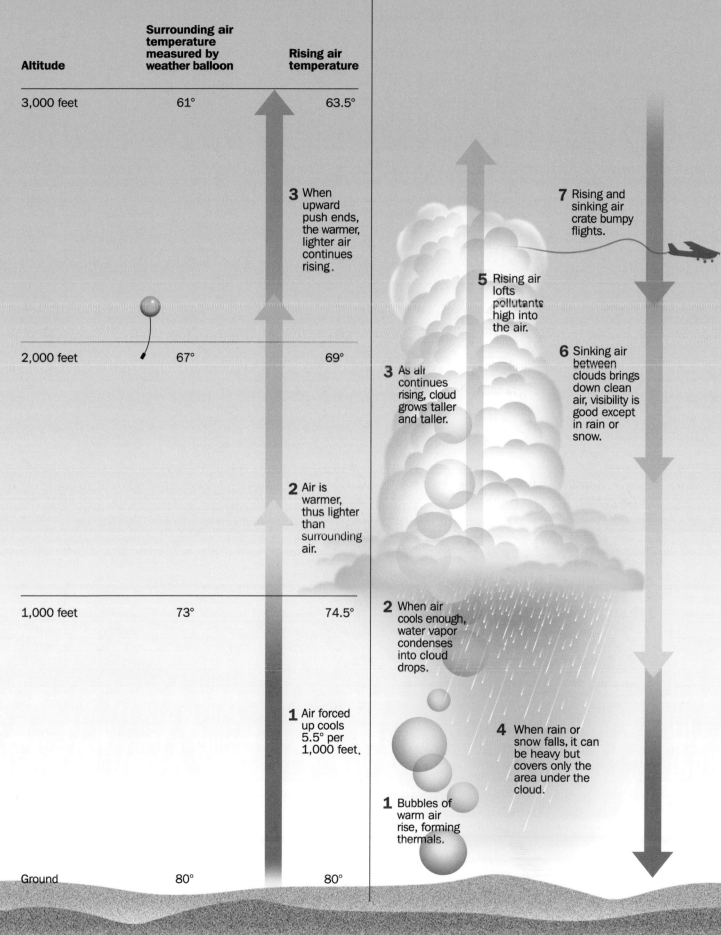

Altitude	Surrounding air temperature measured by weather balloon	Rising air temperature
3,000 feet	61°	63.5°
2,000 feet	67°	69°
1,000 feet	73°	74.5°
Ground	80°	80°

3 When upward push ends, the warmer, lighter air continues rising.

2 Air is warmer, thus lighter than surrounding air.

1 Air forced up cools 5.5° per 1,000 feet.

7 Rising and sinking air crate bumpy flights.

5 Rising air lofts pollutants high into the air.

6 Sinking air between clouds brings down clean air, visibility is good except in rain or snow.

3 As air continues rising, cloud grows taller and taller.

2 When air cools enough, water vapor condenses into cloud drops.

4 When rain or snow falls, it can be heavy but covers only the area under the cloud.

1 Bubbles of warm air rise, forming thermals.

How clouds create rain and snow

A cloud will form or grow where rising air keeps the cloud's tiny water drops or ice crystals from falling. When drops grow too large and heavy for the rising air to hold them up, they fall as rain or snow.

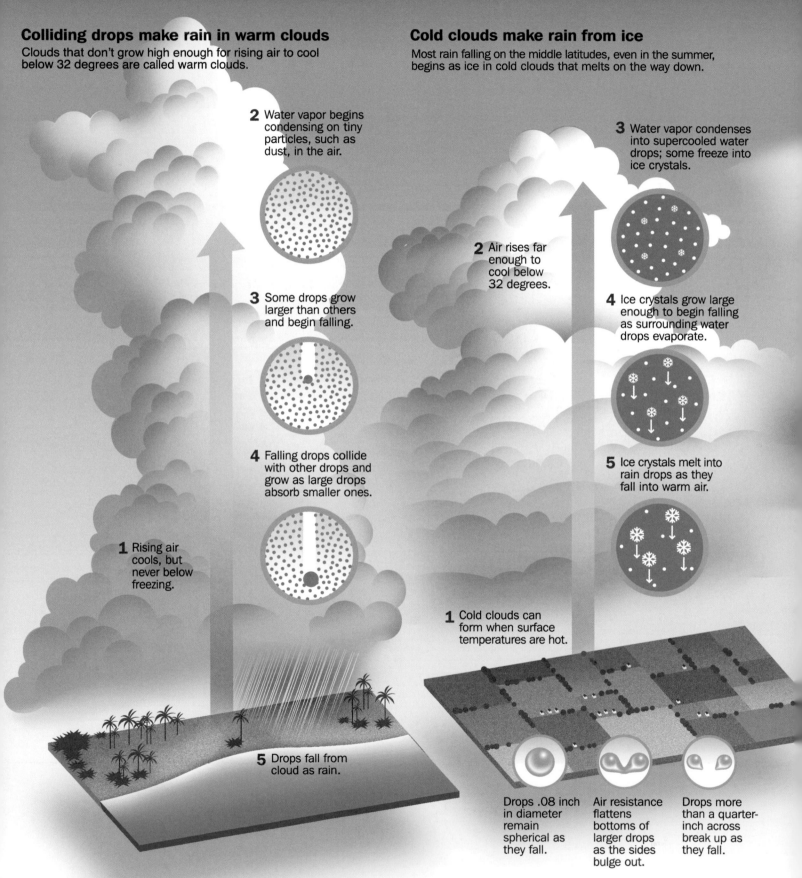

Colliding drops make rain in warm clouds

Clouds that don't grow high enough for rising air to cool below 32 degrees are called warm clouds.

2 Water vapor begins condensing on tiny particles, such as dust, in the air.

3 Some drops grow larger than others and begin falling.

4 Falling drops collide with other drops and grow as large drops absorb smaller ones.

1 Rising air cools, but never below freezing.

5 Drops fall from cloud as rain.

Cold clouds make rain from ice

Most rain falling on the middle latitudes, even in the summer, begins as ice in cold clouds that melts on the way down.

3 Water vapor condenses into supercooled water drops; some freeze into ice crystals.

2 Air rises far enough to cool below 32 degrees.

4 Ice crystals grow large enough to begin falling as surrounding water drops evaporate.

5 Ice crystals melt into rain drops as they fall into warm air.

1 Cold clouds can form when surface temperatures are hot.

Drops .08 inch in diameter remain spherical as they fall.

Air resistance flattens bottoms of larger drops as the sides bulge out.

Drops more than a quarter-inch across break up as they fall.

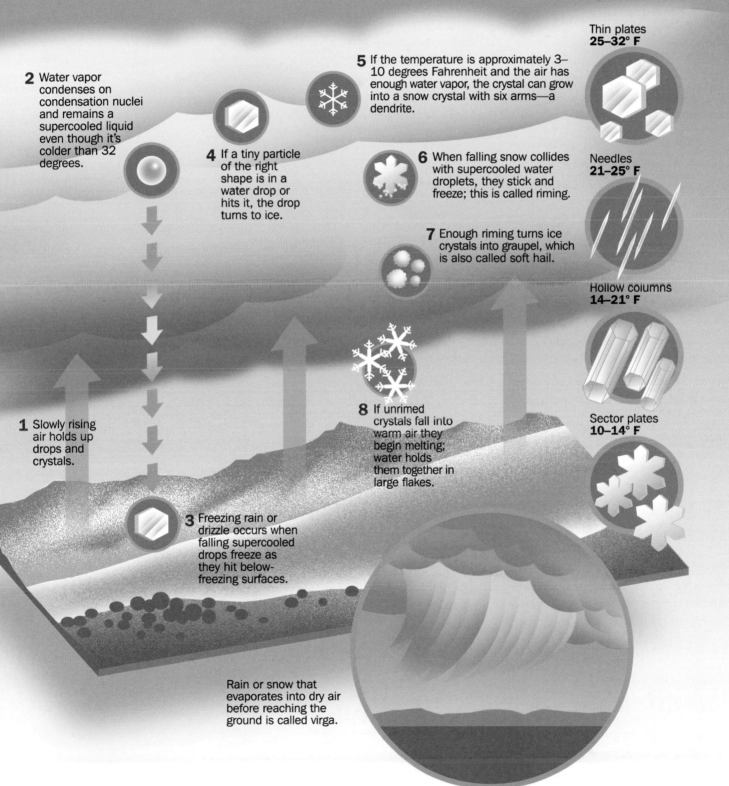

Snow and ice clouds

Snow and other kinds of ice that fall to the ground can grow in different, complex ways.

Temperature determines shapes of snow crystals.

2 Water vapor condenses on condensation nuclei and remains a supercooled liquid even though it's colder than 32 degrees.

4 If a tiny particle of the right shape is in a water drop or hits it, the drop turns to ice.

5 If the temperature is approximately 3–10 degrees Fahrenheit and the air has enough water vapor, the crystal can grow into a snow crystal with six arms—a dendrite.

6 When falling snow collides with supercooled water droplets, they stick and freeze; this is called riming.

7 Enough riming turns ice crystals into graupel, which is also called soft hail.

8 If unrimed crystals fall into warm air they begin melting; water holds them together in large flakes.

1 Slowly rising air holds up drops and crystals.

3 Freezing rain or drizzle occurs when falling supercooled drops freeze as they hit below-freezing surfaces.

Rain or snow that evaporates into dry air before reaching the ground is called virga.

Thin plates
25–32° F

Needles
21–25° F

Hollow columns
14–21° F

Sector plates
10–14° F

Mountains like these in the Transantarctic Mountain range break through the Antarctic Ice Sheet. The large mountain on the right is the 7,054-foot Dufek Massif, named for Rear Admiral George J. Dufek, who commanded the U.S. Navy forces that supported International Geophysical Year activities in 1957–1958.

clouds or even thunderstorms, which are associated with unstable air, embedded in the sheets of stratus clouds.

Oceans and ice

If you were in a spacecraft at least 12,000 miles directly above the Pacific Ocean, halfway between Central America and Australia, you'd see that Earth is a watery planet. Except for approximately half of both Australia and Antarctica, all of New Zealand, and the western edges of North and South America, water is all you would see. Indeed, oceans cover approximately 70 percent of the the earth.

The oceans hold roughly 97 percent of Earth's water. In contrast, only a tiny share of Earth's water is in the atmosphere at any one time, but this tiny fraction is important to weather forecasters, who carefully track it and use this **precipitable water** data in making forecasts. NASA scientists have estimated that if all of the water vapor in the atmo-

sphere condensed and fell as rain, without any more water evaporating, it would cover the entire earth with a little less than an inch of rain.

On a human scale the oceans are deep, averaging more than two miles to the bottom, which means you need to be encased in a vessel that can withstand tremendous pressures to explore most of the ocean floor. This is why so much of the deep oceans remain unexplored. On a planetary scale, however, the oceans really aren't deep. If the Pacific Ocean were the size of an 8½ × 11 inch sheet of paper, its average depth would be approximately the thickness of such a sheet.

Stored water. If you looked at the the earth from a spacecraft at least 12,000 miles above the south-central Pacific in December, during the Southern Hemisphere summer, Antarctica's gleaming ice would surely catch your eye because clean ice reflects 90 percent of the sunlight hitting it. The ice that covers almost all of the Antarctic continent is three miles deep in places and accounts for approximately 90 percent of the the earth's ice.

Glaciers and ice sheets. After Antarctica, the world's next biggest chunk of ice is the 19,000-square-mile **ice sheet** that covers approximately 90 percent of Greenland. The rest of the world's ice consists of **glaciers** across the Northern and Southern hemispheres, including a few atop high mountains in the tropics, such as the 19,335-foot-high Mount Kilimanjaro in Tanzania only 200 miles south of the equator. As we saw above, the oceans hold something in the neighborhood of 97 percent of the world's water, but it's salty. Ice locks up nearly 69 percent of the world's fresh water, at least for now.

Glaciers, **ice caps**, and ice sheets form and grow when the snow that falls during winter doesn't melt the following summer. Obviously, this happens only in places where the temperature seldom rises above freezing at any time of the year. In very cold places you don't expect huge amounts of snow to fall since frigid air isn't going to carry large amounts of water vapor that can turn into snow. Despite what you sometimes hear, it's never really too cold to snow, only too cold for heavy snowfalls. This means that glaciers or ice sheets must have developed over hundreds and thousands of years.

Fresh snow is likely to have a density of less than 5 pounds per cubic foot compared with 62 pounds per cubic foot for fresh water. As snow piles up, it settles. New snow presses down on the old snow, increasing its density. Over time, the snow's crystals begin breaking up and reforming into small, round grains. Eventually the weight of the overlying snow squeezes it enough to make its density approximately 35 to 50 pounds per cubic foot. In this state there is still enough space between the crystals for air to flow through it, and glaciologists call it **firn**. When the increasing pressure has squeezed the crystals so close together that air no longer flows between them, and the density is more than 55 or so pounds per cubic foot, it's considered **glacier ice**. This can take up to 200 years to develop, depending on how much snow falls.

As snow turns into firn and glacier ice, its weight causes it to spread out like a very thick liquid moving slowly across the rock or soil it's sitting on. If the glacier is forming in a mountain valley, as many do, the ice will flow downhill. But even if it's on flat land, the ice will spread out and move. In fact, to be considered a glacier, the ice has to be moving. Ice at the front end of a glacier, especially one that flows downhill into warmer weather, will melt and flow toward the ocean as a stream. If the glacier or ice sheet reaches the water it can push out over the water as floating ice that's still attached to the ice on land, forming an **ice shelf**.

Eventually warmer water or air, and sometimes the motion of the tides and waves, weakens the ice at the end of the ice shelf, allowing pieces to break off to float away as **icebergs**.

As the ice at the end of a glacier or ice sheet melts or breaks off to create icebergs, snow continues to fall, eventually becoming glacier ice. As we will see in Chapter 12, the question about polar ice isn't: "Is polar ice melting?" The question is: "Is polar ice melting faster than new snow is adding to the ice?"

Floods and droughts

The storm the CALJET scientists flew into on February 2, 1998, and again the next day battered California from San Diego north, with winds up to 90 mph and high surf all along the coast. Heavy rain flooded towns, cities, and highways along the coast. Rain falling on the coastal range's eastern slopes sent floods downhill into California's central valley. As the storm moved inland, it spread heavy snow across the mountains of east-central California, with 3 feet of snow falling on the Mammoth Mountain ski area in two days. When snow falls on California's mountains, it's like money in the bank. California, like much of the western United States, depends on mountain snow that melts in the spring and summer to supply water during the dry season that begins in spring and runs into the fall. While

> **Icebergs**
> Since icebergs break off from glaciers or ice sheets, they release freshwater when they melt because they are made of ice that fell as snow.
>
> Like any ice, they float, usually approximately nine-tenths underwater because ice is approximately nine-tenths as dense as water. As we know from the tragedy of the Titanic, the dense ice, mostly hidden under water, can rip apart steel ships.
>
> A few companies and individuals in Canada, Alaska, and Greenland sell chunks of ice or water from icebergs, promoting the purity of water from snow that fell 10,000 or more years ago.

Low-level jets are found in all parts of the world. In Chapter 9, we see how they feed moisture to clusters of overnight thunderstorms.

When high school research turns into a career

The opportunity to conduct real atmospheric research while still in high school put brothers Andrew and Gerald Heymsfield on separate scientific paths that merge from time to time, such as when both were aboard a NASA DC-8 being shaken by 45 mph **updrafts** as it climbed toward the top of Tropical Storm Chantal in August 2001.

Andrew was measuring the sizes, shapes, and amounts of tiny ice crystals and water droplets in the clouds the DC-8 was flying through, continuing research he began as a graduate student in the early 1970s. Gerald was interpreting **Doppler radar** data radioed from NASA's ER-2, flying high above the DC-8. Gerald had begun working with weather radar in graduate school.

Andrew's fascination with ice goes back to early childhood. "When I was three or four years old, I was mesmerized by snow. Whenever there was a forecast of snow I'd get up in the middle of the night and watch it fall."

Meteorology wasn't always on their minds, however. When they were at Brooklyn Tech High School in New York City, Andrew thought about becoming an accountant, because "I was good in math, and I like working with numbers." Automobiles fascinated Gerald, who is two years younger than Andrew. When he was fifteen he and their older brother Steven (now a physician) built a car. "At one point in high school, I was thinking of being an automotive engineer," Gerald says.

Their turn toward atmospheric science began when Andrew was selected for the Natural Sciences Institute, a 1960s and 1970s summer program for high school seniors and college freshmen and sophomores. Following Andrew's advice, Gerald also participated in the Institute, which not only encouraged them to become scientists but prompted both to go to the State University of New York College at Fredonia, one of the sites for the institute's summer programs. Andrew graduated from Fredonia in 1969 with a degree in physics, and Gerald two years later, also with a physics degree.

Both also went to the University of Chicago, where Andrew earned a master's degree in geophysical sciences in 1970 and a PhD in 1973. Gerald earned a master's in geophysical sciences from Chicago in 1972 and a PhD in meteorology from the University of Oklahoma in 1976.

At the University of Chicago, Andrew followed his interest in ice to cirrus clouds, which are made almost completely of ice. Studying cirrus clouds begins with precisely measuring their composition. The university leased for him a World War II-era Lockheed Lodestar—a small, twin-engine transport—so he could fly directly into the object of his research. He describes the airplane as "not very much advanced beyond" the one Helmut Weickmann used for his pioneering cirrus cloud research in Germany in the 1940s (the only such research previously done).

For each flight Andrew dressed for the possible −40-degree cold of the airplane's cabin when it climbed above 20,000 feet…with an open window. Before each flight, the pilots would close the door between the heated cockpit and the cabin and "tell me to have fun," Andrew says. "You do what you have to do to collect data." A tube through the open window brought in ice crystals from clouds; the crystals stuck to glass slides coated with silicone oil. Data, including the time each slide was collected, had to be carefully recorded so researchers would know which part of what cloud it came from. The slides showed crystal shapes, which, among other things, indicate how much solar energy a cloud's ice crystals reflect. Andrew and other scientists measured or calculated data such as the size of the crystals, the amount of water in them, how fast they fell, and how much solar energy they reflected.

In the early 1970s few scientists studied cirrus clouds. "I was about the only one out there for fifteen years doing work on cirrus clouds. It was purely out of curiosity," Andrew says. "[There was] not much known about the properties of cirrus clouds. I was the first one to quantitatively describe them."

Gerald, like most atmospheric scientists of the time, was more interested in severe weather than the pure **cloud microphysics** of phase changes and latent heat transfers on the scale of millimeters or less that Andrew was focusing on. When he arrived at the University of Chicago in 1971, the university was doing pioneering work on Doppler weather radar. "Measuring winds with Doppler radar really captivated me," Gerald says. "I like engineering, and radar has more of an engineering aspect to it as well as pure physics and thermodynamics." After earning his master's degree, Gerald went to the University of Oklahoma in Norman, where he produced one of the first three-dimensional studies of a tornadic thunderstorm using two separate radars—dual

Gerald Heymsfield is standing next to a radar that he helped develop and which is used on NASA's ER-2 high-flying research airplane.

Dopplers—for his PhD dissertation.

After earning his PhD and working for a private meteorological company, Andrew went to the National Center for Atmospheric Research (NCAR) in Boulder, Colorado, in 1975, and he now heads NCAR's Physical Meteorology Group. Gerald became a research meteorologist at NASA's Goddard Space Flight Center in Greenbelt, Maryland, in 1979. He's the principal investigator for two radars on NASA's high-flying ER-2 airplane, including the ER-2 Doppler radar he initiated in the late 1980s.

When Gerald arrived at Goddard, he was researching ways to use weather satellite data to improve severe-storm forecasting, and he "realized that if we had a Doppler radar on a high-altitude aircraft, it would answer some of our questions." This led him to develop a Doppler radar for the ER-2 for studies of air motions in thunderstorms. This radar was later used for investigations related to the joint Japanese-U.S. Tropical Rainfall Measuring Mission (TRMM), the first weather satellite with radar.

The growing concern about climate change has encouraged more scientific interest in clouds because they are part of the puzzle of what will happen as greenhouse gases continue increasing. The growing focus on improving hurricane intensity forecasts has also sparked more interest in cloud microphysics. Ice crystals and water droplets, their phase changes high in a hurricane and the updrafts that carry them, affect a storm's strength. These were some of the things Andrew's cloud instruments and Gerald's radar were measuring in Chantal and the other tropical storms and hurricanes they've flown into.

"What motivates me is to get measurements to improve the science. I think my biggest reward is to do good science with the data," Gerald says. "It takes a lot of work sometimes, but it's worth it. Andy is the same way. We both like new things, new instruments, being on the front edge of technology."

The brothers credit their opportunities to conduct real research when they were in high school and their early college years with leading them into personally satisfying careers that have taken them to places around the world. They would like to see more such opportunities open careers to today's students, allowing them to work on fascinating puzzles while increasing the knowledge needed to address some of society's major challenges.

Unlike hurricanes, droughts in the United States seldom offer dramatic images for television news. In contrast to a hurricane's widespread devastation, an image from the 2005 Midwest drought could have been something like a barge loaded with grain aground on a sandbar that's normally deep under the Mississippi River. The drought was so costly because it affected a part of the United States with some of the most productive farms. In addition to stunting crop growth, the low water in rivers hindered the barges that haul not only grain to market, but also other goods, such as building materials to contractors and stores.

Droughts are deadly in poor parts of the world. In March 2006, for instance, the Save the Children organization reported that "at least three million children, including 600,000 who are under age five, are facing severe food and water shortages in Ethiopia, Kenya, and Somalia," as East Africa's long-term drought continued. The organization said that families were taking such extreme measures as forcing young girls to marry and move out, going two to three days without food, and eating wild plants that could be poisonous.

We might imagine similar scenes in what is now the southwestern United States among the people who lived in the buildings we know as pueblos and in cliff dwellings like those preserved in Mesa Verde National Park in Colorado. During the thirteenth century, the people who lived in these places abandoned their homes. Tree rings and other evidence show a severe drought in the region around this time. Archeologists can't say for sure that drought drove the people out, but it's certainly a possibility.

Scientists who research past climates are sure that over at least the last 1,000 years and probably longer, North America has seen periods of very wet and very dry weather, and we have no reason to think that North America and other parts of the world won't face both again.

Flood forecasts. Despite what the future might bring in terms of wetter or drier weather, people should be concerned with *tomorrow's* forecast for rain or snow, even those who live in desert regions. For example, showers and thunderstorms can cause flash floods to wash down normally dry desert streambeds. Such flash floods regularly kill hikers and campers who didn't learn about the po-

tential dangers of an unfamiliar area. During the West Coast's winter wet season, the precipitation forecast is especially important because many people live in places where water, perhaps combined with mud and rocks, can wash down a hill into their homes. Anyone who lives or works near a stream, even a small one, in any part of the world needs to be at least a little concerned about floods.

Summary and looking ahead

While the 1998 CALJET experiment and those of following years shed light on important aspects of how precipitation comes to the West Coast, its major goal is to find ways to improve precipitation forecasts. For instance, Ralph says, the focus of the 1998 experiments was to correlate **low-level jet stream** winds offshore to the amounts of precipitation falling inland. He said West Coast forecasters had a rule of thumb that a low-level jet of 70 mph or faster should prompt flash flood warnings. But forecasters didn't have good ways to measure these winds over the ocean.

Three weeks after the February 2, 1998, flight, Norman Hoffman, the meteorologist in charge of the National Weather Service's office in Monterey, California, said at a press conference that reports from the WP-3 enabled his office to issue flash flood warnings in the Santa Cruz Mountains on that day and in Santa Barbara the next day, six hours sooner than it otherwise could have.

The concerns of forecasters and public officials go far beyond issuing flash flood warnings for a few locations. Like New Orleans before Katrina, Sacramento, California, and surrounding areas are protected from floods by levees, some of which are in poor condition. In February 2006, California Governor Arnold Schwarzenegger told a press conference, "We are literally today one storm or one big earthquake away from a major disaster." His comments were based on a report by the Sacramento Area Flood Control Agency, which noted that large portions of the 2,300 miles of levees that protect the city and areas around it were built as much as 150 years before and have not been repaired or well maintained.

While a hurricane isn't going to hit this area and while earthquakes cannot be predicted, the Sacramento region has to worry about large floods from rain and melting mountain snow. If mountains and hills around Sacramento are already soaked from winter season rain, a couple of days warning that heavy rain is on the way could save lives and reduce economic losses.

"It's a humbling thing for me," Ralph says. "I know the relevance of our studies. We have a large metropolitan area protected by levees. We've seen what can happen. It makes our abstract concerns more real."

The previous chapters introduced the basics of how the energy that drives the weather arrives from the sun and moves around the earth and how air pressure differences causes the winds to blow. In this chapter, we learned about water's role in the weather. Now we're ready to move to Chapter 5 and learn how energy, winds, and water create the global weather patterns that determine what happens today, tomorrow, and far into the future.

Inside El Niño, monsoons, extra-tropical storms, jet streams, and other major weather makers

CHAPTER 5

Even though she was aboard a ship that had just reached the equator, 2,730 miles south of her sixth-grade classroom at Emory Elementary School in San Diego, Dana Tomlinson wasn't taking a vacation from teaching. Instead, on that day, March 11, 2002, she was giving her students and others an on-the-scene look at an **El Niño's** birth.

"What an interesting day, all the way around," she wrote in that day's Web diary from the National Oceanic and Atmospheric Administration (NOAA) research ship *Ka'Imimoana* (its name is the Hawaiian word for "ocean seeker") "Weather-wise, we awoke to clear skies, with clouds on the horizon, and we could tell it was going to be hot. By 9 a.m, I could feel the backs of my legs burning with my back to the sun. I went in for lunch and came out and it was totally clouded over and a few minutes later, it was raining! Not drizzling, raining. Welcome to the equatorial Pacific."

Preceding pages: Kevin Kinsey, left, and Linda Stratton, from the NOAA ship Ka'Imimoana in the background, work on one of the Pacific Ocean buoys that track El Niño.

That day's noontime shower could have been one of the subtle signs of the nascent El Niño that she shared with students via the Web during her three weeks on the ship. In a summary shortly before leaving the ship, Tomlinson said, "The scientists saw the signatures of El Niño: warmer than **normal** sea surface temperatures by 1 degree and a rainfall pattern that has shifted southward and south of the equator."

As one of the few teachers that NOAA invites to take part in research cruises, Tomlinson was not only at sea teaching science via the Web but also collecting experiences and developing concepts and enhanced appreciation for science that she continues to share with her classes and other teachers. "I feel more comfortable teaching science after being out there doing it myself," she said four years later. Before working with scientists and technicians on the ship, Tomlinson says, "I didn't have a sense of scientists as part of a team working together. You are like the gears that make the machine run. This has changed me as a teacher." Since returning from the cruise, "I have gone out of my way to work as a team with my fellow teachers."

The men and women aboard the *Ka'Imimoana* maintain an array of fifty-five ocean buoys strung more than 5,000 miles across the tropical Pacific Ocean on both sides of the equator from 105 degrees west longitude (less than 1,000 miles west of the Galapagos Islands) to the International Date Line. Japan tends to a dozen such buoys from the Date Line west toward Indonesia. The buoys' ocean and atmospheric data help meteorologists and oceanographers predict and study **ENSO**.

While Tomlinson was aboard, the *Ka'Imimoana* visited all seven buoys along the 110 degrees west longitude line, a north-south distance of 897 miles. The ship's crew hauled out and replaced four of the buoys for repairs, made minor repairs to one in the water, and found the remaining two in good condition.

The hands-on science that Tomlinson did aboard the *Ka'Imimoana* included helping Raye Foster, a graduate student who was collecting data for researchers. One day she helped Foster scrape barnacles from one of the recovered buoys, then sort and bag them to be sent to Cynthia Venn at Bloomsburg University in Pennsylvania, who studies the effects of El Niño and **La Niña** on barnacles. Tomlinson also helped launch weather balloons and assisted with deployment of a CTD—the conductivity-temperature-depth device that oceanographers use to study the ocean's depths. Often, as

ENSO

Since the 1980s, many scientists and forecasters have used ENSO, the acronym for El Niño-Southern Oscillation, to refer to the entire cycle with a warm, a neutral, and a cold phase.

We examine the entire phenomenon in this chapter, including the **Southern Oscillation** itself. In this and other chapters, we use ENSO to refer to the entire pattern, El Niño for ENSO's warm phase and La Niña for the cold phase, as researchers and forecasters do. The term ENSO-neutral or ENSO-neutral state are also sometimes used.

on the *Ka'Imimoana*, a CTD is attached to a frame that holds fourteen cylinders, each 4 or 5 quarts in volume, which can be closed and sealed when they reach a particular depth, providing scientists with samples of the water and anything in it.

In another diary note a few days before the ship arrived in the Galapagos Islands, where Tomlinson left to fly home, she wrote, "We are seeing more animal life: two huge pods of porpoises and a few different kinds of birds. Seeing the birds is nice. We have seen very few on this trip. Dr. Michael J. McPhaden, the NOAA chief scientist aboard, feels this could also be an indicator of El Niño since the waters are warmer, the fish may be fewer and, therefore, the birds have less to eat."

Tomlinson's experiences aboard the *Ka'Imimoana* and continuing involvement in using oceanography and meteorology in her classes led the American Meteorological Society to select her for its Maury Project, which prepares master teachers to help their colleagues teach oceanography. In July 2005, she was one of the twenty-four teachers who spent two weeks studying in the

Water's electrical conductivity gives a precise measurement of salinity, which can be used with temperature and depth to calculate density, an important driver of ocean currents, as we saw in Chapter 3.

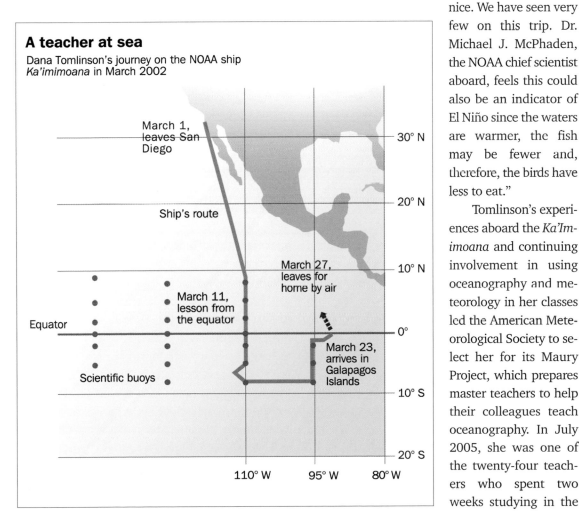

A teacher at sea

Dana Tomlinson's journey on the NOAA ship *Ka'imimoana* in March 2002

March 1, leaves San Diego

Ship's route

March 11, lesson from the equator

Equator

Scientific buoys

March 27, leaves for home by air

March 23, arrives in Galapagos Islands

30° N
20° N
10° N
0°
10° S
20° S

110° W   95° W   80° W

Dana Tomlinson works with students in a reading class at Emory Elementary School in San Diego.

tion, including the jet streams that help steer storms. Changes in large areas of high and low air pressure and in storm paths alter precipitation patterns, bringing drought to some places and floods to others as far away as Africa.

To help you understand El Niño and its effects, we will first look at the global circulation of the ocean and atmosphere before turning to El Niño and other ocean-atmosphere patterns that affect the weather.

oceanographic classrooms and laboratories, and aboard research craft at the U.S. Naval Academy in Annapolis, Maryland.

"That was a very intense experience, especially for someone like me who doesn't have a science background, because we were being taught by PhDs," she says. In Annapolis, "I made sure I sat next to a high school advanced placement physics teacher. When I was having difficulty, I turned to him and said, 'Can you give this to me one more time?'l" With that training and Maury Project support, she now offers workshops for teachers on using oceanography, including where to find scientifically sound materials to use.

Tomlinson has some advice for other teachers like her who don't have a scientific background: "If you are an intelligent adult who's been through college and working as a professional, you can stretch that part of your brain, especially if you are with people who are willing to help you."

El Niño

El Niño is not a single event like a hurricane or a winter storm. It involves a complex, interconnected set of changes in ocean temperatures, winds, and currents across the tropical Pacific that affects the strengths and locations of areas of high and low atmospheric pressure above the Pacific. These changes, in turn, affect global atmospheric circula-

The weather you experience on any particular day depends on atmospheric motions ranging in scale from global to local to turbulence taking place in areas just fractions of an inch across. For example, the sun's unequal heating of the earth explains why Alaska is cooler than Florida. That plus Earth's general circulation of winds and ocean currents are the reasons why hurricanes come from the ocean to hit Florida but not Alaska.

Patterns that modify the global circulation, such as the El Niño pattern that those aboard the *Ka'Imimoana* were experiencing in 2002, can tilt the odds toward a warm winter in Alaska while at the same time increase the chances for a wet winter in Florida. However, the ever-changing locations and intensities of areas of high and low air pressure determine the paths of storms that hit both states. Small-scale weather phenomena and geography can cause a winter storm to bury Anchorage, Alaska, under 2 feet of snow while dumping only 3 inches on Elmendorf Air Force Base just north of the city. This is one reason why scientists stress they can't point to patterns such as El Niño or long-range climate changes (such as those related to global warming) as the "cause" of a particular weather event such as a winter storm that slams California or a heat wave on the East Coast. But patterns such as an El Niño or long-range climate changes can make some kinds of weather events more likely.

Physical and political science

A common but often unspoken assumption in meteorology is that all forecasts are valuable. Michael H. Glantz found early in his career at the National Center for Atmospheric Research (NCAR) that this isn't always the case.

He had earned a master's degree in 1963 and a PhD in 1970, both in political science from the University of Pennsylvania, after switching from a budding career in metallurgical engineering, his undergraduate major when he graduated from Penn in 1961. "It was good…I liked it," he says of his engineering work at Westinghouse and at Ford. But "there wasn't enough people contact…some of us are happy filling a box; I wanted to fill a matrix."

Glantz came to NCAR in 1974 as a **postdoc** to study the societal effects of using cloud seeding to reduce hail damage in the western states. When that idea died, he was asked by NCAR's founder and former director, Walter Orr Roberts, to look into the value of any kind of long-term forecast. Since Glantz had been interested in African drought, he conducted a study of the potential benefits of a perfect drought forecast for Africa's Sahel, the region south of the Sahara Desert that had been devastated by five consecutive drought and famine years. His conclusion was that even a very good forecast would have little value because the poor nations affected couldn't do anything to respond due to the lack of roads, storage facilities, and the like. A similar study for Canada showed little value because the nation and its farmers were already prepared to cope with drought by changing the types of crops.

In the meanwhile, "I had learned that every once in a while an El Niño comes and devastates the rich fisheries along Peru's Pacific coast. If we can forecast that, there's value there." J. Dana Thompson, an oceanographer and fellow postdoc at NCAR, teamed up with Glantz in soliciting a collection of articles about El Niño. Glantz's boss rejected their proposal because "El Niño has nothing to do with the atmosphere." At this time, only a few atmospheric scientists seemed to understand El Niño or even to care about its origins and worldwide impacts. Glantz and Thompson did the book anyway, but their editor told them, "No one will know what El Niño is, don't use it in the title." John Wiley & Sons Inc. published the book in 1981 as

Michael Glantz says other social scientists should consider how climate, water, and weather are connected with their work.

Resource Management And Environmental Uncertainty: Lessons from Coastal Upwelling Fisheries, a year before the 1982–1983 El Niño became news.

Glantz went on to do three more books on El Niño while also continuing to work on the societal effects of African drought and other topics. He continued "questioning the value of meteorological information by itself, without putting in the social context."

In the 1970s, companies that wanted to learn more about El Niño and El Niño forecasts began contacting him. After the 1982–1983 event, governments saw the need to fund El Niño research and monitoring systems. In 1997–1998, "the people's El Niño" encouraged much hype and many jokes but also increased awareness by millions of people around the globe that climate shifts could affect them.

In 2003, Glantz published the book *Climate Affairs: A Primer.* "I used the term affairs to get away from an over focus on climate policy and climate science. You can't do climate assessments without looking at science, impacts, policy, law, economics, ethics, and equity." In fact, Glantz thinks university programs must include philosophers to help students grapple with issues of ethics and equity.

"Science by itself is like clapping with one hand," Glantz says. "You need to have the other side to make a sound, the social side." Forecasts and other scientific results should no longer "be served up by scientists like a smorgasbord where they lay out stuff and the smart people in society will figure out how to use it. You need the translators."

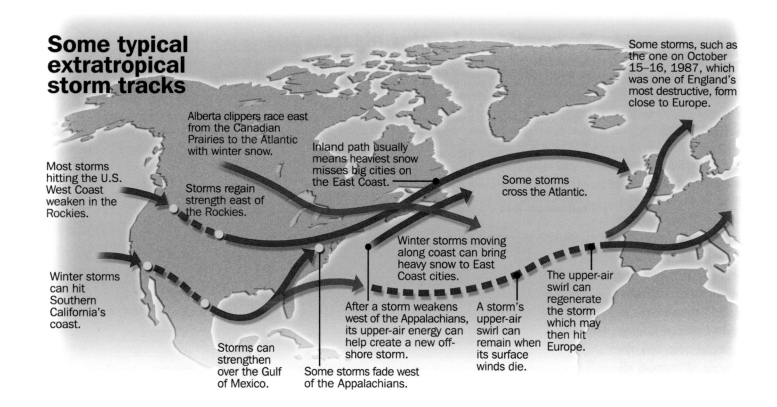

Some typical extratropical storm tracks

Most storms hitting the U.S. West Coast weaken in the Rockies.

Winter storms can hit Southern California's coast.

Alberta clippers race east from the Canadian Prairies to the Atlantic with winter snow.

Storms regain strength east of the Rockies.

Inland path usually means heaviest snow misses big cities on the East Coast.

Some storms cross the Atlantic.

Some storms, such as the one on October 15–16, 1987, which was one of England's most destructive, form close to Europe.

Winter storms moving along coast can bring heavy snow to East Coast cities.

The upper-air swirl can regenerate the storm which may then hit Europe.

Storms can strengthen over the Gulf of Mexico.

Some storms fade west of the Appalachians.

After a storm weakens west of the Appalachians, its upper-air energy can help create a new off-shore storm.

A storm's upper-air swirl can remain when its surface winds die.

June through October, accounting for 61 percent of the yearly precipitation. In contrast, Mumbai's four wettest months bring 95 percent of the yearly rain.

Rising air over the searing deserts of northern Mexico and the U.S. Southwest creates a thermal low that helps cause winds to blow into the area from the Gulf of California and even over the mountains from the Pacific Ocean. The summer expansion of the Bermuda high helps create east-to-west winds that push humid air from the Gulf of Mexico over high terrain into northern Mexico and the U.S. Southwest.

The need to better understand monsoons and their year-to-year differences has driven climate research since the nineteenth century. In fact, as we see below, monsoon research helped scientists begin putting together the climate puzzle we now know as El Niño.

A fruitful model

Even if you've paid little attention to the weather until now, you have likely encountered meteorology's most common model of complex events. For example, you've seen weather maps on television, on the Internet, or in a newspaper that include the letters "H" and "L" (for areas of high and low air pressure, respectively), blue lines studded with triangles (cold fronts), and red lines with half circles (warm fronts). You might know, for example, that centers of low air pressure tend to be cloudy, and if a cold front is heading your way, temperatures could drop.

These maps reflect the model that scientists led by the Norwegian physicist Vilhelm Bjerknes working in Bergen, Norway, developed around the time of World War I. It's much more than a good way to depict middle-latitude weather on maps, it represents an important step toward making the study of weather a mathematical science firmly grounded in physics. Bjerknes, like many other meteorologists of his time, was convinced that the mathematical laws of physics could be used to predict the weather. Meteorologists finally realized this vision after World War II, building on work of earlier scientists including Bjerknes, as we will see in Chapter 7.

The Bergen meteorologists started using the term front in 1919 as an analogy to the battlefronts that had been a big part of the news during World War I, which had ended the year before. Front is appropriate because, as Bjerknes said in a talk at the British Royal Meteorological Society in 1920, "We have before us a struggle between a warm and

A conveyor belt storm model

The image below shows details of some extratropical cyclones, including those that bring heavy snow to the U.S. northeast. It helps explain aspects of satellite images such as the large "comma cloud" shown as white in the image. (Based on Paul J. Kocin and Louis W. Uccellini, *Northeast Snowstorms*, Volume 1, Overview, Boston, American Meteorological Society, 2004, p. 129)

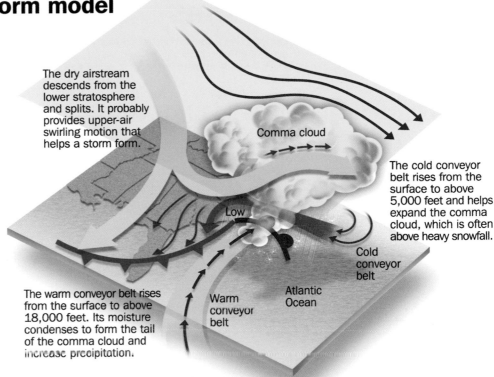

The dry airstream descends from the lower stratosphere and splits. It probably provides upper-air swirling motion that helps a storm form.

Comma cloud

The cold conveyor belt rises from the surface to above 5,000 feet and helps expand the comma cloud, which is often above heavy snowfall.

Low

Cold conveyor belt

Atlantic Ocean

Warm conveyor belt

The warm conveyor belt rises from the surface to above 18,000 feet. Its moisture condenses to form the tail of the comma cloud and increase precipitation.

a cold air current. The warm [air] is victorious to the east of the centre. Here it rises up over the cold, and approaches in this way a step toward its goal, the pole. The cold air, which is pressed hard, escapes to the west, in order suddenly to make a sharp turn towards the south, and attacks the warm air in the flank: it penetrates under it as a cold West wind."

The storms the Norwegian model describes are **extratropical cyclones**, which form outside of the tropics. Contrasts between warm and cold air supply almost all of the energy for these storms. Extratropical cyclones form and grow over land or oceans, including over chilly oceans. As we see in Chapter 10, tropical cyclones, which include hurricanes and **typhoons**, form and grow only over warm oceans and weaken when they move over land or cold water.

The Norwegians described extratropical storms as forming along the "polar front," which divides cold air around the poles from warm tropical and subtropical air. In theory, this wavy front stretches around the globe in both the northern and southern hemispheres, with cold air advancing toward the tropics in some places, and warm air pushing toward the poles in others. In winter, the entire front is closer to the tropics than in the summer when it

retreats back toward the pole. We say, "in theory," because weather charts rarely show a continuous string of fronts stretching around a hemisphere. Even during winter, north-to-south changes in temperatures, pressure, and winds are too gradual in some areas for meteorologists to locate a front.

In three dimensions

The Norwegian model of middle-latitude storms is three dimensional, with air rising along frontal zones, but the meteorologists in Bergen, like their colleagues elsewhere, had little data to help them confirm or reject hypotheses about the weather aloft. By the 1930s, however, meteorologists were obtaining increasing amounts of upper-air data, which helped improve their understanding of weather aloft. This included understanding why bands of high-speed winds called jet streams form and the role they play in the atmosphere's global circulation and in day-to-day weather. The two-page graphic in Chapter 3 shows how pressure gradients aloft and the Coriolis force create jet streams.

Going back to the early twentieth century, a few meteorologists using telescopes to track small balloons as they rose to high altitudes had measured winds faster than 150 mph. In March 1935,

Middle latitude storms

The extratropical cyclones that form over land or oceans in the middle latitudes are responsible for major blizzards and fierce ocean storms as well as for the ever-changing weather in the latitudes between the tropics and polar regions. The images on these pages are based on the model originally developed by meteorologists working in Bergen, Norway, early in the twentieth century and refined since then.

An extratropical cyclone's birth, life, and death

1 Setting the stage

A cold, high-pressure mass of air moves in from the northwest.

A warm, high-pressure mass of air arrives from the southeast.

Since neither air mass is displacing the other, the boundary between them is not moving, it's a stationary front

2 A wave forms

A wave forms on the stationary front.

Cold air begins flowing under less-dense warm air.

Warm air begins flowing over denser cold air.

Fronts show a boundary between air masses at the surface

Close-up of a cold front

Rising warm, humid air creates cumulus clouds.

Winds behind the front usually blow from the west or northwest. Weather offices use the shift in wind direction to mark passage of cold fronts.

Arriving cold air pushes warm air up.

Frontal movement

Southerly surface winds ahead of the front

Inside an occluded front

Unlike stationary fronts, cold fronts, and warm fronts, which separate two air masses, an occluded front separates three air masses.

Widespread clouds and precipitation

Warm air

Boundary at the ground

Very cold air

Cold air

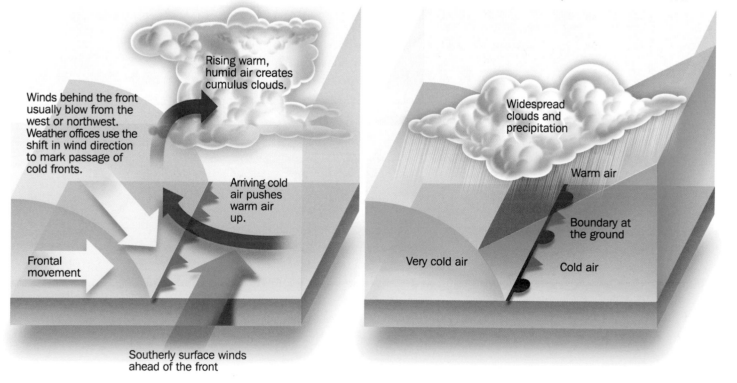

3 The storm forms and grows

The entire system is moving toward the east as the pressure falls in the low-pressure center.

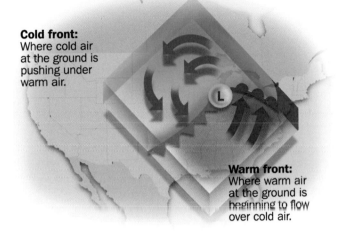

Cold front:
Where cold air at the ground is pushing under warm air.

Warm front:
Where warm air at the ground is beginning to flow over cold air.

4 The storm begins to die

The low-pressure center redevelops in the cold air north of the cold and warm fronts. The storm weakens and dies as air pressure in the center rises.

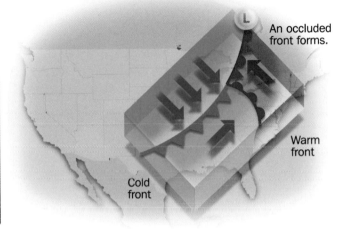

An occluded front forms.

Warm front

Cold front

What happens in a warm front

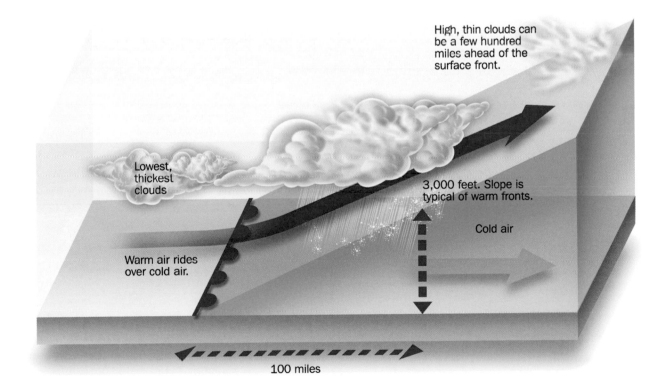

High, thin clouds can be a few hundred miles ahead of the surface front.

Lowest, thickest clouds

3,000 feet. Slope is typical of warm fronts.

Cold air

Warm air rides over cold air.

100 miles

Waves in upper-atmospheric winds

The jet stream illustrated below has well-defined Rossby waves. They are called long waves because only two to five are needed to circle the hemisphere.

Trough of cold air with low atmospheric pressure aloft.

Ridge of warm air with high atmospheric pressure aloft.

Winds in a trough can impart a spinning motion to the air called an **upper-level disturbance** that often causes clouds and precipitation.

Jet stream

A short wave, which is a ripple in a long wave, moves along the jet stream.

Locations and temperatures of cold and warm air, Earth's rotation, and high mountain ranges influence Rossby wave patterns, which in turn affect movements of warm and cold air.

Cutoff lows

A short wave, with its spinning air motion, is moving toward the east with the upper-air flow.

Cold air and low pressure move toward the south, becoming pinched off from the upper-air flow.

Once cut off from the flow, the spinning pool of cold, low-pressure air can stay in place or even drift west with its clouds and precipitation.

Winds aloft and surface pressure

Changes in the speeds or directions of upper-air winds can cause the air to flow together (**convergence**) in some places and to flow apart (**divergence**) in other locations.

1 Convergence occurs where airflow slows or changes from clockwise to counterclockwise on the east side of a ridge.

4 Divergence occurs where airflow speeds up or changes from counterclockwise to clockwise, such as on the east side of a trough.

2 Convergence pushes air down.

5 Divergence encourages air below to rise.

3 Descending air forms high atmospheric pressure at the surface.

6 Rising air creates low pressure at the surface.

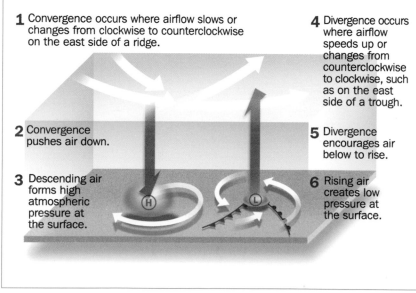

the American aviator Wiley Post flew his single-engine Lockheed Vega from Burbank, California, to Cleveland, Ohio, as high as 35,000 feet in a record seven hours and nineteen minutes, pushed by winds averaging at least 100 mph and at times reaching at least 160 mph.

Citing this and his other high-altitude flights, some reference works list Post as the discoverer of the jet stream. But Post, like other 1930s aviators, was thinking more in terms of going faster in the thinner air aloft than in catching rides on the winds. In fact, in an assessment of Post's flight, the *New York Times* science writer Russell Owens wrote on March 24, 1925, that high-altitude fliers such as Post had discovered that "up to 15,000 or 20,000 feet, wind of great velocity, usually from the west, may be encountered" but above those altitudes winds blow from any direction and "there are no storms and no bumps, only masses of air moving at moderate speed." Unfortunately for airline passengers who are uncomfortable with turbulence, those early aviators weren't quite correct about "no bumps" at high altitudes, but there certainly is much less turbulence than at lower altitudes where piston-engine airliners such as the DC-3 flew. Other reference works with more of a meteorological bent list the Swedish-American meteorologist Carl-Gustaf Rossby (1898–1957) as the discoverer of the jet stream. This claim has a stronger foundation but does not represent the whole story.

Rossby is best known today for working out the mathematics of the wavy paths that upper air winds follow around the globe. These upper-atmosphere long waves, which are now called **Rossby waves**, not only move heat toward the poles and cold air toward the tropics, but also help form areas of high and low air pressure at the surface and steer storms.

Rossby's many accomplishments include heading the World War II program that provided university training for more than 6,000 meteorologists in the armed forces. As part of this work, he and other academic meteorologists traveled to U.S. Army and Navy forecasting offices in Europe and the Pacific, where they heard stories from bomber and fighter pilots who encountered high-altitude winds faster than 150 mph.

After the war ended in 1945, the U.S. military funded upper-atmosphere research at universities that, among many other things, laid the founda-

Storm energy

A simple experiment using a fish tank with an easily removable partition dividing it demonstrates how air density difference between cold air and warm air supplies energy to drive extratropical cyclones.

1 Fill one side with warm, fresh water marked with red food dye.

2 Fill the other side with very salty, cold water marked with blue food dye.

3 Pull out the partition.

4 Heavier, cold water flows down and under the warm water…

5 …pushing it up.

Similar density differences between air masses occur along fronts.

tions for understanding how pressure patterns drive the jet stream and the role of jet streams in the general circulation.

Meteorologists were quick to share their growing understanding of upper-air circulation and its effects on weather with the public. On September 6, 1953, a *New York Times* story under the headline, "'Jet stream' Flowing Out of Siberia Plays A Part in Our Hot and Cold Spells" by Waldemar Kaempffert, the *Times* science and engineering editor, describes the jet stream as being at least partly responsible for "last week's hot weather."

The story reflects the fact that by 1953, the leading meteorologists of the time were familiar not only with 200-plus mph jet stream winds at alti-

tudes of 30,000 to 50,000 feet but also appreciated their role in day-to-day weather. This was leading-edge science at the time. Kaempffert describes the forecasters who used the jet stream to predict the hot spell as "the modernists among meteorologists." He cites the work of Rossby, Hurd C. Willett (1903–1992) of the Massachusetts Institute of Technology (MIT), and Jerome Namias (1910–1997), who was then chief of the Weather Bureau's long-range forecasting division.

Today you often hear statements such as "the jet stream has plunged into the South, bringing cold air." But that's an over simplification. High-altitude jet streams flow atop strong boundaries between warm and cold air. High-altitude jet streams help determine the paths of extratropical cyclones, which in turn move cold air toward the tropics and warm air toward the poles. Jet streams and masses of cold and warm air engage in a dance, with neither the undisputed leader.

Atmospheric rivers

The NOAA scientists on the research flight described in the opening of Chapter 4 were studying a storm being fed by tropical water vapor from hundreds of miles to the southwest when they encountered the low-level barrier jet. In addition to documenting what happens to humid, tropical air when it reaches the U.S. West Coast, the scientists using data from both their flights and from satellites confirmed a hypothesis of how humid tropical air reaches the middle latitudes.

Since at least the 1930s, atmospheric scientists thought that relatively narrow bands of strong, low-level winds, sometimes called low-level jets, haul in water vapor to feed middle-latitude storms. This hypothesis was finally confirmed beginning in the 1990s when two MIT researchers, Yong Zhu and Reginald E. Newell, examined global weather data with a new computer model and concluded that three to five narrow rivers of air carry approximately 90 percent of the tropical water vapor reaching the middle latitudes at any one time. In a 2004 scientific journal paper, F. Martin Ralph, Paul Neiman, and Gary Wick of NOAA's Environmental Technology Laboratory in Boulder, Colorado, wrote that their study, using both flight and satellite data, "confirms the model-based conclusions presented by Zhu and Newell."

West Coast meteorologists sometimes refer to an atmospheric river as a **pineapple express**. The term originally referred to streams of moisture from near Hawaii but is often used for any atmospheric river.

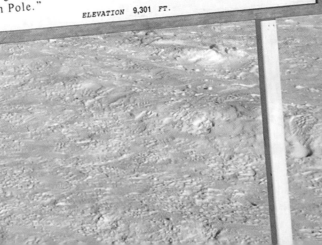

GEOGRAPHIC
SOUTH POLE

ROALD AMUNDSEN   ROBERT F. SCOTT

DECEMBER 14, 1911   JANUARY 17, 1912

"So we arrived and   "The Pole. Yes, but
were able to plant our   under very different
flag at the geographical   circumstances from
South Pole."   those expected."

ELEVATION 9,301 FT.

How people and instruments keep track of what's going on in the atmosphere and oceans

CHAPTER 6

Shelley Knuth, George Weidner, John Kominko, and Chuck Slade worked hard all one afternoon in 2006 to set up an automated weather station in Antarctica. Their work is paying off. Since they set up the station it has been helping forecasters make predictions and helping researchers learn a little more about Antarctica's climate as far back as 100,000 years ago.

At the time, Knuth was a graduate student in meteorology at the University of Wisconsin–Madison and Weidner a researcher at the University's Antarctic Meteorological Research Center. They worked out of the U.S. McMurdo Station in Antarctica from late December 2005 through January 2006, servicing and setting up automated weather stations.

Kominko and Slade were the pilots who flew the DeHavilland Twin Otter that carried Knuth and Weidner to stations on the ice sheet during the second week of January 2006.

Previous pages: The U.S. Amundsen-Scott South Pole Station is an important part of the global weather and climate observing system. In the lower right, the silver pole with the brass plaque on top marks the geographic South Pole. The red and white pole with the reflective globe on top is the ceremonial South Pole. The building is the new South Pole Station, which was dedicated January 15, 2008.

With no other flights scheduled that day, the pilots volunteered to spend the afternoon of January 13 helping Knuth and Weidner set up the new station, only a short snowmobile ride from the new U.S. West Antarctic Ice Sheet Divide field camp, where they were staying a few days.

The automated station, like others around Antarctica, is designed to continue radioing temperature, humidity, wind speed and direction, and air pressure data to satellites for several years, including during winters when temperatures regularly drop below −60 Fahrenheit and no one is around to make repairs.

Shelley Knuth works on an automated station in Antarctica.

Data from the automated stations help forecasters predict storms that would endanger airplanes flying to and from remote camps. They are also the primary source of data that will show whether the continent's climate is changing.

On January 13, 2006, the four began their work by making three trips using a 1980s snowmobile pulling sleds loaded with tower parts and tools from the camp to the location selected for the station about half a mile from the camp. "Putting up the station wasn't too difficult," Knuth said in an e-mail to friends the next day. "We first dug a three-foot hole [in the snow] to get the tower in, attached the instruments, and then raised it. It was a great accomplishment."

With the weather tower up, the four turned to digging a three-foot-deep pit in the packed snow and then drilling down more than twenty feet to install a string of sensors that are tracking ice temperatures for researchers.

Using their muscles to power the drill "was quite taxing" Knuth said in her e-mail. They had to pull the drill out every few feet to clear ice from it. "The drill itself was as tall as the hole," she wrote, "so it was quite hard to pull it out towards the end. The pilots were a great help, and to reward them, George and I decided to name the station after them."

With the tower erected and the digging and drilling in the snow finished, "George and I were spent," Knuth wrote in her e-mail report. After returning to the camp, "we wired a couple more batteries for the next day, had dinner (grilled pork chops), and went to sleep. By the way, the camp had two cooks, and the food was really good. They always had fresh bakery [goods] available. The morning we left, for breakfast we had fresh cinnamon rolls."

National Science Foundation Antarctic field camps have a few buildings, usually canvas huts that are relatively easy to erect and take down. These supply comfortable places to cook, eat, work, and socialize.

Those working at such camps, however, often sleep in tents, usually mountain tents, as Knuth did. "The temperature was in the upper teens to the lower 20s there so it was a bit chilly to sleep in," she wrote to her friends. But the heat of one person in a small tent warms it during the night, and "it was all right by the end of the night," she wrote.

Knuth reminded her friends that in January Antarctica has twenty-four hours of sunlight each day. "So it was pretty bright when I was trying to sleep." A sleep mask solved that problem, and "I was comfortable when I fell asleep." But during her second night in the tent, unlike the first, "there were no clouds, so the sun was really bright and my tent got really, really warm by morning. It must have been over 80 degrees in there, so I didn't sleep very well."

In addition to working on the automated

weather stations, Knuth dug pits deep enough to go through two years of snow accumulation at several locations to measure the snow's density. This gave her an indication of whether the snow was light and fluffy, which means the wind would blow it around more than heavy snow.

She was collecting this information for her master's degree, which she earned in 2007 with a thesis entitled "Estimation of Snow Accumulation in Antarctica Using Automated Acoustic Depth Gauge Measurements." Ordinary snow measurements don't distinguish between snow that drifted to a location from snow that fell there, making measuring snowfall tricky. How much snow falls in Antarctica "is a missing piece of the puzzle of the global water cycle," Knuth says. Solving the puzzle "is another key to understanding the global climate."

In addition to taking part in an important scientific enterprise, Knuth and others who go to Antarctica as researchers or support staff see places that tourists pay dearly to visit. When someone asks Knuth what she paid to go to Antarctica, she answers: "I was paid to go there."

On July 3, 2006, when the nearest people were 750 miles away at the year-round South Pole station, the Kominko-Slade station recorded a temperature of −73.3 degrees Fahrenheit and has continued to report through winter's darkness and

Workers lounge in the galley tent during their lunch break at the West Antarctica Ice Sheet Divide field camp.

Tracking Antarctica's weather

Shelley Knuth at one of the automated weather stations she worked on in Antarctica. The box she's leaning against was approximately eight feet above the surface of the snow when the station was installed. She's standing on packed snow that has accumulated since then. The stations often operate for years without anyone visiting to service then and shovel away snow.

Measuring temperature

Radiation shield blocks direct sunlight and sunlight reflected from snow to prevent heating the temperature probe.

The open top and bottom of the shield allow air to flow freely.

3"

3"

6"

Antenna sends data to satellites for relay to the University of Wisconsin's Antarctic Meteorological Research Center.

Box holds the barometer, which is an electronic device that senses air pressure, and the computer that collects data from all sensors and prepares messages to transmit.

Temperature probe contains a platinum wire that's part of an electrical circuit. The wire's electrical resistance changes with the temperature of the air around it. Measurements of the circuit's resistance show the temperature.

The 1911 instrument shelter in the foreground is typical of those still used globally to shield thermometers from direct sunlight. It's at Cape Evans on McMurdo Sound in Antarctica. The building in the background was the home base for Robert Falcon Scott's 1911–1912 expedition. George Simpson, the party's meteorologist, used readings from here and elsewhere for his three-volume treatise on Antarctic Weather.

the twenty-four hours of daylight during the late November through February summer when scientists, technicians, and those who run the camp are there drilling through the ice and conducting other research.

Weather observations

Weather-reporting stations are vital because meteorologists need to know what the atmosphere and ocean are doing in real time to forecast what will happen in the future. In Chapter 7 we will see how meteorologists use this data for this purpose. Solid information about the state of the atmosphere and oceans in the past, sometimes the very distant past, is vital for scientists trying to predict Earth's future climates.

Antarctica's stations. Forecasters trying to predict quickly forming storms, which could endanger airplanes flying to places such as the West Antarctic Ice Sheet Divide field camp, rely on much less data than meteorologists anywhere else in the world. Upper-air data come from balloons launched from only a dozen weather stations around the coast plus reports from the few airplanes flying over Antarctica. Weather satellites don't collect as much data from the Antarctic or the Arctic as from the rest of the world, and Antarctica has no weather radar. The entire continent has approximately twenty manned stations reporting surface weather plus the approximately 200 automatic stations operated by the United States and other nations.

In contrast, in the contiguous U.S. forty-eight states forecasters receive hourly observations from more than 11,000 surface weather stations (mostly automated). Airliners flying over the forty-eight states send approximately 35,000 automated, high-altitude observations each day. Balloons launched from fifty weather balloon stations twice a day add more high-altitude data. Finally, a network of Doppler weather radars give meteorologists detailed views of the weather over virtually all of the forty-eight states.

Sometimes the Antarctic stations operate for years with no one around, like one that Kominko and Slade flew Knuth and Weidner to the day before they put up the new station. Knuth says it was still operating even though no one had visited in eight years and drifting snow had covered all but the wind-measuring anemometer, and the boom that holds the temperature and relative humidity sensors, "so those were still reporting." She says. "The solar panel was also still above the surface so that was able to keep the station powered." In this case, they bolted new tower sections on to those buried in the snow to raise the instruments to their regular height above the snow.

Up to a dozen twelve-volt, gel-cell batteries charged by one or two ten-watt solar panels power the stations. Stations at the South Pole, where winter darkness lasts just a few days short of six months each year, need twelve batteries and two solar cells, while stations farther north, where winter darkness is shorter and temperatures are warmer, need fewer batteries and solar cells. The stations transmit data every ten minutes to the joint French–U.S. Argos data collection system on U.S. polar orbiting satellites, which sends it on to ground stations.

The West Antarctic Ice Sheet camp weather station named for the pilots is especially vital from late November into February as meteorologists forecast the weather for airplanes transporting supplies, equipment, and personnel to and from the camp. Those working at the camp are drilling through the 2.1 miles of ice under the camp to extract climate data. Even though forecasts aren't needed during the eight months each year when the camp is closed, the data are added to Antarctica's sparse climate record to aid scientists tracking whether Antarctica is warming or cooling.

Drilling for data

While weather stations around the world, including Antarctica's automated stations, are helping track what the climate is doing now, glaciers and ice sheets are a storehouse of data about past climates. In Chapter 4, we saw that when snow falls in places too cold for it to melt in the summer, it piles up to form glaciers and ice sheets. As more and more snow is added, the snow on top squeezes the older snow below, eventually turning it into glacier ice. After World War II, scientists decided to see whether ice from deep inside glaciers or ice sheets could tell them anything about past climates. If nothing else, they thought, air bubbles and dust trapped for years, probably centuries, could be valuable.

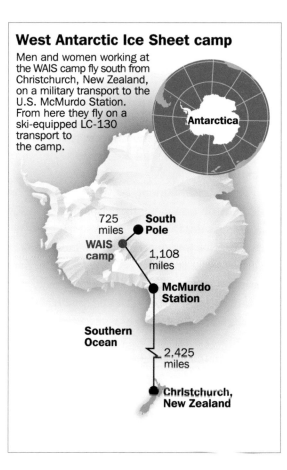

West Antarctic Ice Sheet camp

Men and women working at the WAIS camp fly south from Christchurch, New Zealand, on a military transport to the U.S. McMurdo Station. From here they fly on a ski-equipped LC-130 transport to the camp.

Antarctica

725 miles — South Pole

WAIS camp

1,108 miles

McMurdo Station

Southern Ocean

2,425 miles

Christchurch, New Zealand

The Argos system was developed in the 1970s to collect data from ocean buoys. Today's version can pick up signals from transmitters weighing less than an ounce, like those scientists attach to migratory birds.

As it turns out, even more data than air bubbles or dust is locked into glacier ice. The very atoms of oxygen and hydrogen in the ice's water molecules are a good measure of air temperatures when snow fell. Water molecules aren't all alike because both oxygen and hydrogen have more than one **isotope**. In nature, approximately 99.8 percent of all oxygen atoms are oxygen-16; they have a nucleus made of eight protons and eight neutrons, giving them an atomic mass number of 16. Oxygen-18 is also found in nature, accounting for almost all of the remaining 0.2 percent of oxygen atoms. It has a nucleus also made of eight protons, but 10 neutrons, giving these oxygen atoms a mass number of 18. Hydrogen also has isotopes, with by far the most common being hydrogen-1, with one proton in the nucleus. Hydrogen-2 (called deuterium) has one proton and one neutron.

Almost all water molecules are made of one atom of oxygen-16 and two atoms of hydrogen-1 for a total molecular mass of 18. But a water molecule made of one oxygen-18 atom and two hydrogen-1 atoms will have a molecular mass of 20. A water molecule with a molecular mass of 18 and another with a mass of 20 act the same chemically.

Measuring the oceans' depths

Approximately 3,000 Argo floats send regular ocean data on water temperature and salinity from the surface to more than a mile deep. Oceanographers, climate scientists, fisheries managers, weather forecasters, and others use the data, which now come from all of the world's ice-free oceans.

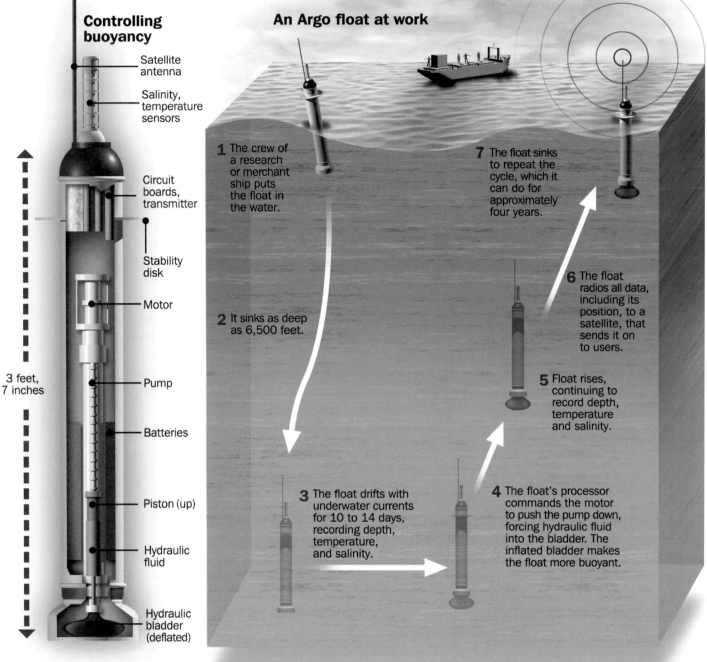

Controlling buoyancy

- Satellite antenna
- Salinity, temperature sensors
- Circuit boards, transmitter
- Stability disk
- Motor
- Pump
- Batteries
- Piston (up)
- Hydraulic fluid
- Hydraulic bladder (deflated)

3 feet, 7 inches

An Argo float at work

1 The crew of a research or merchant ship puts the float in the water.

2 It sinks as deep as 6,500 feet.

3 The float drifts with underwater currents for 10 to 14 days, recording depth, temperature, and salinity.

4 The float's processor commands the motor to push the pump down, forcing hydraulic fluid into the bladder. The inflated bladder makes the float more buoyant.

5 Float rises, continuing to record depth, temperature and salinity.

6 The float radios all data, including its position, to a satellite, that sends it on to users.

7 The float sinks to repeat the cycle, which it can do for approximately four years.

the Netherlands, Argentina, and the United States worked from 2000 through 2003 compiling a database of wind reports from British, Dutch, French, and Spanish naval vessels. Researchers are using the data to track patterns of phenomena such as El Niño during periods when no one realized they existed. The century ending in 1854 is important as it

marks the end of the era during which humans likely did not affect global climate by adding large amounts of greenhouse gases to the air.

Nineteenth-century naval officers logged wind data because the wind powered their ships. Extracting weather data from these logbooks requires translating the specialized vocabulary of eighteenth- and early nineteenth-century sailors and doing calculations to correct for navigational errors made long ago by ships' officers.

Ocean data

As we saw in Chapters 2 and 5, the oceans are an integral part of Earth's climate system. To forecast the weather, even for places far inland, meteorologists need data on ocean temperatures, waves, and currents. Tracking the changes in the global patterns that we looked at in Chapter 5 requires ocean data.

The 1982–1983 El Niño, which was the strongest in more than one hundred years, brutally made this point by catching forecasters completely off guard. Not only did they not predict that an El Niño was forming and what its effects might be, they even didn't realize it was going on until months after it started.

By 1982 researchers had learned much about El Niño, but that year's big event didn't begin in the way they had come to expect. Changes in the patterns of equatorial Pacific Ocean sea surface temperatures are one of the most important signs that an El Niño is underway, but satellite readings of Pacific Ocean temperatures were misleading, as researchers later discovered. Eruptions of El Chichon volcano in Mexico in March and April 1982 had spread a thin veil of tiny **aerosols** around the world, which threw off satellite measurements of ocean temperatures without anyone realizing it until the El Niño was well underway.

At the time, the United Nations' World Climate Research Programme was planning a large, joint study called the Tropical Ocean-Global Atmosphere (TOGA) project. As Michael Glantz says in his book, *Currents of Change: El Niño's Impact on Climate and Society*, the 1982–1983 "event gave a strong impetus and a sense of urgency to the need to carry out TOGA." The experiment, which lasted from 1985 to 1994, left a legacy of not only more knowledge about El Niño but also the array of Tropical Ocean Atmosphere (TOA) buoys across the tropical Pacific. The TOA buoys are moored, that is they are anchored to the ocean bottom. (In the story that opens Chapter 5, Diane Tomlinson, a teacher aboard a NOAA ship, is visiting some of these buoys.) Other kinds of buoys are anchored elsewhere in the oceans, especially along coasts, to automatically monitor both the weather and oceans.

Since the year 2000, the United States and several other nations have been deploying buoys that NOAA refers to as "robot oceanographers." These Argo floats perform a job traditionally handled by people on oceanographic research ships, measuring the temperature, salinity, and movement of ocean water both at the surface and below. While shipboard measurements have taught scientists much about the oceans, they can no longer supply the vast amounts of data forecasters and researchers need.

Once a ship or airplane drops an Argo float into the ocean, it sinks, drifts with the current for ten days, and then comes to the surface to transmit the data, usually to the Argos receiver on a NOAA satellite. These satellites also collect data from Antarctic weather stations, migrating birds and whales, and hundreds of other sources.

Surface data

Meteorologists forecasting U.S. weather use data from 1,000 NWS, military, and other government weather offices. In addition, more than 10,000 automated weather stations (some government, many private) send regular reports.

Ed Johnson, director of the NWS Office of Strategic Planning and Policy, described the surface weather observing system as "almost amazingly complex…because ownership is so dispersed and many institutions own surface observations in one way or another. If you own a network you install it for your purposes. You own the data and you get to choose who gets it and on what terms."

If a federal agency collects the data, there's no question about making those data available to everyone, Johnson says. Since taxpayer money pays for producing the information, anyone who wants it has to pay only the delivery cost, not the collection costs. Almost all of the U.S. weather data you see on television, in newspapers, or on Web sites are collected by the NWS or are from NWS offices, the U.S. military, the Federal Aviation Administration (FAA), and other government agencies, or has been transmitted via NOAA satellites.

However, some of the data come from private companies that make their readings generally available. By the summer of 2008, one of the largest of these, WeatherBug, had 8,000 professional-grade weather stations and more than 1,000 cameras at

In Greek mythology, Argo was the ship Jason and the Argonauts sailed on their search for the Golden Fleece. Argo floats sail the twenty-first century seas and Argo is teamed with a satellite called JASON-1 that precisely measures global sea levels.

Automated weather observations

Automated weather stations use devices made of materials with electrical properties affected by temperature, pressure, or humidity to measure these weather elements. Measuring precipitation amounts, the kinds of precipitation, visibility, and wind speed and direction is more difficult to automate. The NWS Automated Surface Weather Observing System (ASOS) uses the instruments shown below.

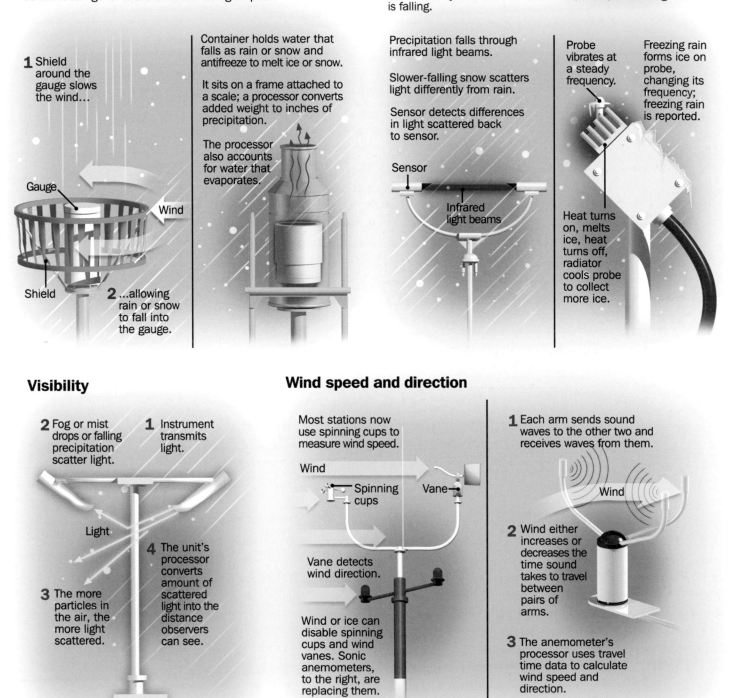

Precipitation measurements

The National Weather Service is replacing older kinds of rain gauges with the all-weather precipitation accumulation gauge below. It can go for months without being emptied.

1 Shield around the gauge slows the wind...

Gauge

Wind

Shield

2 ...allowing rain or snow to fall into the gauge.

Container holds water that falls as rain or snow and antifreeze to melt ice or snow.

It sits on a frame attached to a scale; a processor converts added weight to inches of precipitation.

The processor also accounts for water that evaporates.

Precipitation type

Gauges measure the total inches (or millimeters) of rain, drizzle, and melted ice and snow as precipitation. More complex devices automatically determine whether rain, snow, or freezing rain is falling.

Precipitation falls through infrared light beams.

Slower-falling snow scatters light differently from rain.

Sensor detects differences in light scattered back to sensor.

Sensor

Infrared light beams

Probe vibrates at a steady frequency.

Freezing rain forms ice on probe, changing its frequency; freezing rain is reported.

Heat turns on, melts ice, heat turns off, radiator cools probe to collect more ice.

Visibility

2 Fog or mist drops or falling precipitation scatter light.

1 Instrument transmits light.

Light

3 The more particles in the air, the more light scattered.

4 The unit's processor converts amount of scattered light into the distance observers can see.

Wind speed and direction

Most stations now use spinning cups to measure wind speed.

Wind

Spinning cups

Vane

Vane detects wind direction.

Wind or ice can disable spinning cups and wind vanes. Sonic anemometers, to the right, are replacing them.

1 Each arm sends sound waves to the other two and receives waves from them.

Wind

2 Wind either increases or decreases the time sound takes to travel between pairs of arms.

3 The anemometer's processor uses travel time data to calculate wind speed and direction.

Measurements of cloud cover

The National Weather Service uses terms such as "partly cloudy" with specific meanings when it reports how much of the sky clouds actually cover or are forecast to cover, as seen from the ground. Different terms are used in reports and forecasts for pilots. Terms used for the public and for pilots are given in the illustrations below. When no clouds are seen, the sky is "clear."

- Public: Mostly clear (night) or mostly sunny(day)
- Pilots: Few clouds
- Clouds cover 1/8 to 2/8 of the sky

- Public: Partly cloudy
- Pilots: Scattered clouds
- Clouds cover 3/8 to 4/8 of the sky

- Public: Mostly cloudy or considerable cloudiness.
- Pilots: Broken clouds
- Clouds cover 5/8 to 7/8 of the sky

- Public and pilots: Overcast
- Clouds cover all of the sky

For pilots, the distance between the ground and the bottoms of broken or overcast clouds constitutes the ceiling.

Automated cloud measurements

Pilots need information on actual and forecast cloud heights as well as cloud cover. Almost all measurements today are by automated ceilometers, which measure cloud cover and how far the bottoms of clouds are above the ground.

4 Upper-air winds push clouds across the sky. Every minute the ceilometer uses the past 30 minutes of data on when clouds were overhead to calculate cloud cover.

2 If light hits a cloud, some of it is scattered back toward the ceilometer.

1 A ceilometer sends pulses of infrared laser light straight up.

3 The ceilometer uses the time needed for light to make the round trip (at the speed of light) to calculate cloud height.

The NWS is replacing older ceilometers that can measure clouds up to 12,000 feet above the ground with ones that can measure clouds up to 25,000 feet.

schools, public buildings such as fire stations, and television stations around the United States automatically reporting conditions each second. The American Meteorological Society (AMS), publisher of this book, is one such participant, with data-collecting equipment installed on top of its Beacon Hill headquarters in Boston.

The NWS supports the large MesoWest network, which is a cooperative project between University of Utah researchers; Salt Lake City National Weather Service Office forecasters; the NWS Western Region Headquarters; and men and women working for other agencies, universities, and com-

mercial firms. The goal is to provide current weather observations in the western United States, where the mountains can help create large weather differences over short distances. From its founding in 1997, the network has grown to more than 10,000 stations.

An increasing amount of weather data is also coming from road information networks operated by state, local, and private highway operators. Johnson says that perhaps thousands of organizations of various kinds are collecting surface environmental observations. Since many of these rely on NOAA satellites to transmit the data, the infor-

Into the heart of hurricanes

When Nicole Mitchell graduated from high school, her mother and stepfather, who were both in the National Guard, suggested the Guard would be a good way to pay for college.

"At first I was a little resistant," she says, but the Minnesota Air National Guard's available jobs included weather forecasting. "I was always into math, I loved science, and I loved the outdoors. Weather fit. I didn't think of using it as a career, but it just stuck."

Nicole Mitchell is a Weather Channel meteorologist who also flies with the Air Force Reserve Hurricane Hunters, where she has one big advantage: she doesn't suffer from air sickness.

In addition to helping her pay for college, Guard service led Mitchell into a dual career as an on-camera meteorologist at the Weather Channel in Atlanta, and as an Air Force Reserve weather officer with the 53rd Weather Reconnaissance Squadron based in Biloxi, Mississippi.

In the military she received an education equivalent to that of an undergraduate meteorology major, and she worked as a forecaster in Europe, Saudi Arabia, and the United States.

Mitchell began her civilian career at a Duluth, Minnesota, television station "where you dabble at everything…reporting…a little producing" and on-air weather, which led to other television jobs.

While completing her Air Force degree in weather, she was also earning a University of Minnesota bachelor's degree in speech-communications, which qualified her for an officer's commission.

Mitchell transferred to the Hurricane Hunters in 2003, and her first flights were into winter storms, which the 53rd also flies into. Her first tropical storm was Charley over the Caribbean Sea in August 2004, when forecasters thought—based on satellite estimates—that its maximum winds were only 45 mph. The WC-130 airplane flew into Charley as low as 500 feet above the ocean. "It was more intense than anyone thought. We observed hurricane-force surface winds in the northeast quadrant—far greater than expected. You usually don't want to be at a low level in an intense storm. I could see the ocean swirling, which was fascinating."

On storm flights, Mitchell is the flight meteorologist and mission director. "I kind of direct where plane is going to go, but the pilot is the aircraft commander" with a veto on anything dangerous.

Airplanes fly through hurricanes from side to side across the eye, usually at least 5,000 feet above the ocean, several times during an average mission, which typically lasts ten hours but can exceed thirteen hours. Usually an airplane encounters strong turbulence for only a few minutes each time it flies through the most intense winds around the eye. But Hurricane Emily in July 2005 was an exception. "We had turbulence most of the storm. After five or six hours of turbulence I was just exhausted, my body was saying, enough of this," Mitchell says.

At times Mitchell is a little torn between her two jobs, especially in 2005 when she flew into four hurricanes and three tropical storms. "That year was insane…I was driving back and forth almost every weekend between Atlanta and Biloxi to go fly. If there is a landfalling storm, that's when I need to be flying. At the same time, the Weather Channel needs everyone on deck." However, the Weather Channel has enough on-camera meteorologists that it can spare her for storm flying.

"It's a fascinating job," Mitchell says. "I think for anyone who's a meteorologist and has a sense of adventure, which I definitely do, it's kind of a perfect melding of science and adventure."

mation becomes available to anyone who wants to use it. In addition to ordinary weather readings, such as temperatures and winds, these data include water levels in streams, readings of the dryness of woodland areas that could catch fire, and soil temperatures and moisture amounts.

Looking aloft

From the beginning of the rise of modern meteorology, scientists have wanted to learn about the air above the earth's surface. Consider the account in Chapter 3, when in 1648 Pascal asked his brother-in-law to take a barometer to the top of a mountain to see whether air pressure decreases with altitude. In 1783, the brothers Joseph and Etienne Mongolfier opened the sky to atmospheric scientists and their instruments when they made the world's first hot air balloon flights. Scientists have also used kites, beginning with Benjamin Franklin and his famous 1752 lightning experiment. Realizing the risk, few people launched kites into thunderstorms, but kites are handy for carrying instruments, such as thermometers, aloft. In fact, a few researchers continue to use kites because they can be easily sent aloft to study lower altitudes, carrying lightweight instruments or cameras, and cost little to use.

By late in the nineteenth century, meteorologists knew that upper-air observations could improve forecasts, and in 1898 the U.S. Weather Bureau began regular kite observations. The kites were nothing like the kites children fly on a breezy day. The Weather Bureau's box kites were more than six feet tall and hooked to strong, thin, piano wire wrapped around a steam-driven reel. They regularly flew higher than 10,000 feet, with additional kites every 2,000 feet along the wire to supply more lifting force.

In 1925 the Weather Bureau began paying pilots to carry weather instruments aloft and ended kite flights in 1933. The Weather Bureau didn't pay pilots unless they climbed to at least 13,500 feet, which probably encouraged some to take chances; crashes killed twelve weather observation pilots in the 1930s. By the middle of the 1930s, weather services around the world began sending balloons aloft carrying **radiosondes**, which are packages of instruments combined with radio transmitters that can send data back. In 1940, weather balloons car-

Meteorological time

All nations normally give the times of weather observations and forecasts in Coordinated Universal Time (UTC), which avoids the need to convert among different local times. Weather services also use the 24-hour (or military) clock, with 1 p.m. becoming 1300, 2 p.m. 1400, and so forth until midnight, which is 2400. One minute past midnight is 0001, 1 a.m. is 0100, and so on.

UTC is commonly referred to as "Zulu" time because a letter identifies each time zone, and the international radiotelephony spelling alphabet uses "Zulu" for the letter "Z," the time zone at Greenwich, England. When the United Kingdom or other parts of the world use daylight time, UTC (Zulu) time does not change.

This is why a radar image on the U.S. National Weather Service Web site from 7 p.m. Eastern Time will show 00:00 Z in the winter, but 23:00 Z in the summer, when the United States is on daylight time.

UTC time got its name in 1970 when the International Telecommunication Union adopted the concept of Coordinated Universal Time, and the French wanted to use "TUC" for "temps universel coordonné," while English speakers wanted to use "CUT" for "Coordinated Universal Time." The compromise was "UTC."

rying radiosondes grounded the last of the U.S. Weather Bureau's airplanes that had been collecting what are known as upper-air soundings.

Today, approximately 800 weather stations around the globe, mostly in the Northern Hemisphere, routinely launch balloons carrying radiosondes twice a day at noon and midnight Coordinated Universal Time. (Many poor nations often launch only one balloon each day.) The balloons rise and expand as air pressure decreases, finally bursting at approximately 115,000 feet above sea level. A small parachute lowers the shoebox-size radiosondes to Earth, where most are lost. The U.S. NWS says people find and return less than twenty percent of the approximately 75,000 radiosondes it releases each year. If you ever find a radiosonde, you'll see instructions on how to return it printed on its side.

Automated airliner data. With only 800 or so balloons being launched globally twice a day, radiosondes miss a lot of important weather. Weather services and aviation authorities have long encouraged pilots to radio in reports about the weather they are flying through; these pilot reports help forecasters keep up with what's going on aloft. A problem is that pilots tend not to report weather

For many purposes, UTC is the same as Greenwich Mean Time (GMT), named for Greenwich, England, outside London. UTC is based on atomic clocks and GMT is based on astronomical time.

Doppler weather radar

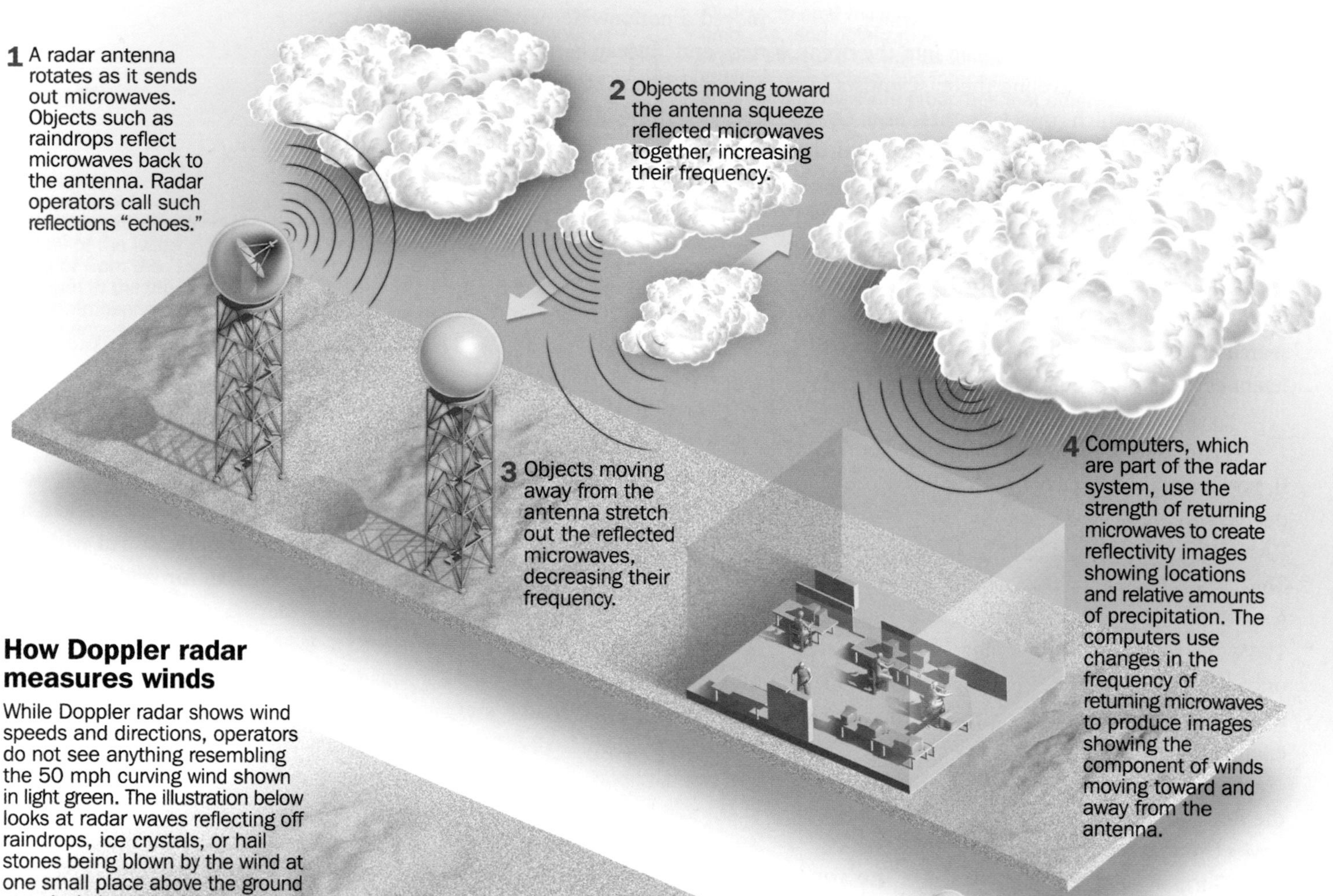

1 A radar antenna rotates as it sends out microwaves. Objects such as raindrops reflect microwaves back to the antenna. Radar operators call such reflections "echoes."

2 Objects moving toward the antenna squeeze reflected microwaves together, increasing their frequency.

3 Objects moving away from the antenna stretch out the reflected microwaves, decreasing their frequency.

4 Computers, which are part of the radar system, use the strength of returning microwaves to create reflectivity images showing locations and relative amounts of precipitation. The computers use changes in the frequency of returning microwaves to produce images showing the component of winds moving toward and away from the antenna.

How Doppler radar measures winds

While Doppler radar shows wind speeds and directions, operators do not see anything resembling the 50 mph curving wind shown in light green. The illustration below looks at radar waves reflecting off raindrops, ice crystals, or hail stones being blown by the wind at one small place above the ground at a single instant. Millions of such measurements create Doppler images like the one to the right.

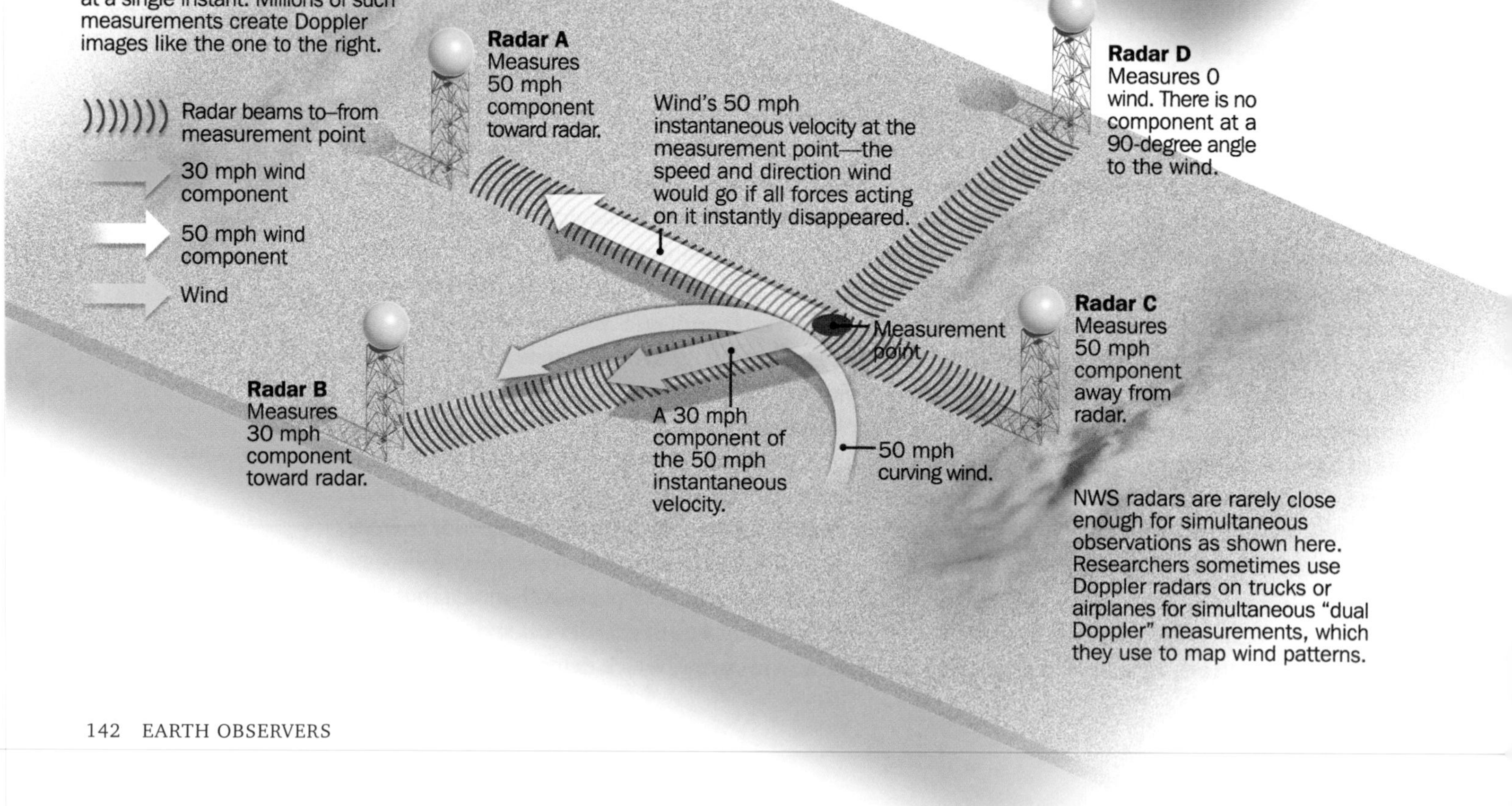

))))))) Radar beams to–from measurement point

→ 30 mph wind component

⇒ 50 mph wind component

→ Wind

Radar A Measures 50 mph component toward radar.

Wind's 50 mph instantaneous velocity at the measurement point—the speed and direction wind would go if all forces acting on it instantly disappeared.

Radar D Measures 0 wind. There is no component at a 90-degree angle to the wind.

Radar C Measures 50 mph component away from radar.

←Measurement point

Radar B Measures 30 mph component toward radar.

A 30 mph component of the 50 mph instantaneous velocity.

50 mph curving wind.

NWS radars are rarely close enough for simultaneous observations as shown here. Researchers sometimes use Doppler radars on trucks or airplanes for simultaneous "dual Doppler" measurements, which they use to map wind patterns.

The images here show the reflectivity and Doppler images from the Lincoln, Illinois, NWS Doppler radar at 2:43 p.m. on July 13, 2004, as a strong tornado was destroying the Parsons Company, Inc., factory near Roanoke, 46 miles to the north. Chapter 8 opens with the tornado story and later describes the Lincoln office's forecasts.

Reflectivity

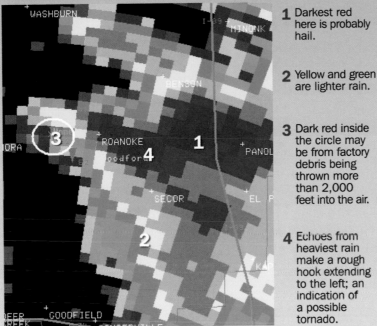

1 Darkest red here is probably hail.

2 Yellow and green are lighter rain.

3 Dark red inside the circle may be from factory debris being thrown more than 2,000 feet into the air.

4 Echoes from heaviest rain make a rough hook extending to the left; an indication of a possible tornado.

Almost all "Doppler" images seen on television show reflectivity.

Doppler winds

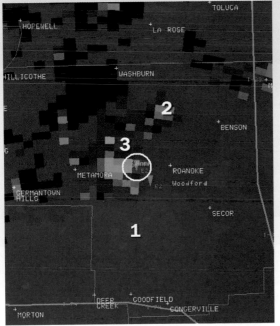

1 Brightest reds are fastest winds away from the radar.

2 Brightest greens are fastest winds toward the radar.

3 Adjacent bright green and bright red squares indicate winds faster than 58 mph in opposite directions next to each other.

This image shows storm-relative motion; the winds after the motion of the storm is subtracted. In other words, the winds as they would be if the storm were not moving.

direction an airplane was moving across the ground or ocean.

The Doppler shift, of course, affects all radar waves, but earlier, ordinary radars didn't have the processing ability needed to measure and display Doppler information. By the middle of the 1970s, the NWS was looking for a replacement for its 1957 and 1974 model radars, which used hard-to-replace vacuum tubes. The Air Force Air Weather Service and the Federal Aviation Administration were also looking for a new weather radar. During the tornado seasons of 1977 through 1979, these agencies conducted a joint study based in Norman, Oklahoma, which led to the decision to develop the Next Generation Weather Radar, or NEXRAD. It is also called the WSR-88D radar. "WSR" stands for Weather surveillance radar, the "88" indicates that the prototype was built in 1988, and the "D" means it's Doppler.

The WSR-88D's computers turn raw radar data into a variety of images that help give forecasters a better handle on the weather. They are also more sensitive, able to detect phenomena such as bands of heavy snow in big storms that the old

Polarized weather radar

Radar scientists and engineers at the National Severe Storms Laboratory in Norman, Oklahoma, and at other places have demonstrated that polarizing Doppler weather radar microwaves and measuring the polarization of reflected microwaves provides more information than ordinary Doppler radar.

Polarized radar microwaves

Horizontal and vertical polarization of radar microwaves refers to the orientation of the electrical field. The microwaves shown below are propagating horizontally like the microwaves represented in the images to the right.

Horizontal polarization

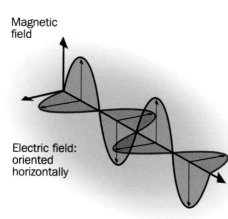

Magnetic field

Electric field: oriented horizontally

Vertical polarization

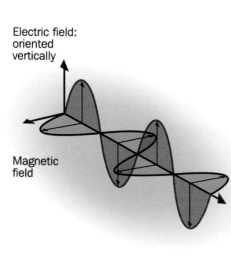

Electric field: oriented vertically

Magnetic field

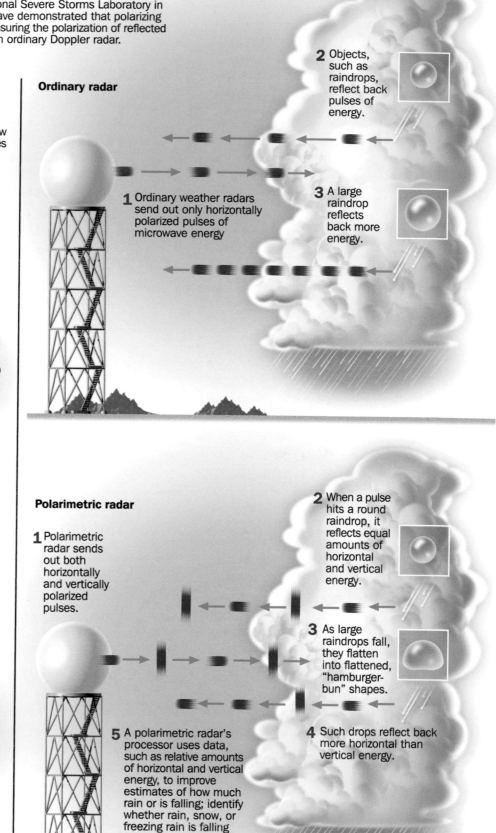

Ordinary radar

2 Objects, such as raindrops, reflect back pulses of energy.

1 Ordinary weather radars send out only horizontally polarized pulses of microwave energy

3 A large raindrop reflects back more energy.

Polarimetric radar

1 Polarimetric radar sends out both horizontally and vertically polarized pulses.

2 When a pulse hits a round raindrop, it reflects equal amounts of horizontal and vertical energy.

3 As large raindrops fall, they flatten into flattened, "hamburger-bun" shapes.

4 Such drops reflect back more horizontal than vertical energy.

5 A polarimetric radar's processor uses data, such as relative amounts of horizontal and vertical energy, to improve estimates of how much rain or is falling; identify whether rain, snow, or freezing rain is falling from winter clouds; and determine the size of hail stones.

radars saw as a blur. Between 1991 and 1998, the new radars replaced all of the outdated weather radars in the United States. The new radars included methods for sending data to private companies that can enhance it in various ways for television stations, Web sites, and other users. Radar images on television could be based on data from the network of WSR-88D radars or from a Doppler radar manufactured for private meteorologists, including those affiliated with television stations.

Twenty-first century radar. Just as you see television stations today promoting their Doppler weather radar, in the next few years or so you may see them promoting something like "Polarimetric Doppler 2015" and, in a decade or so, the "Phased Array 2020 Radar." A polarimetric radar can do a much better job of distinguishing between falling rain, hail, and snow and this can be very important.

The graphic on the adjacent page illustrates the basic science of polarimetric radar and a few of its uses. While Doppler radar has revolutionized forecasting and research, says John Snow, dean of geosciences at the University of Oklahoma, it has shortcomings that adding polarization capability will not solve.

Antennas of today's weather radars, such as the WSR-88D, spin around to send their beams out in all directions. When forecasters want to concentrate on a particular area, such as a thunderstorm, they can have the antenna focus on that area, probably tilting the antenna up and down for a complete look. Even so, Snow says, "Thunderstorms can move 60 mph, and most move around 40 mph." One of today's radars "takes six minutes to do a scan. We don't see the thunderstorm, we see a smeared-out picture." Instead of using a spinning antenna to point its beam, a phased-array radar directs the beam electronically, which means, "we will see what happens in a 45-degree segment in a minute. We will get away from time-smearing."

Phased array radar was developed by the U.S. Navy, which has to worry about detecting and shooting down missiles zooming just above the waves toward its ships. A phased array radar more quickly detects a missile and supplies data to the ship's fire control computers, which aim the ship's rapid-firing guns or anti-missile missiles at the threat. Meteorologists are interested in phased-array radar for a similar reason, it can more quickly detect the shifting winds inside a thunderstorm associated with formation of a tornado. In 2000, the Navy turned over a phased array radar to the Severe Storms Lab, and researchers there are working to turn it into a weather radar.

James F. Kimpel, Director of the Severe Storms Laboratory, says phased array radar represents an even bigger step into the future than the WSR-88D did in its day. The Lab and the FAA are looking into the possibility that the same phased array radar could be used to watch the weather over a wide area, focus on individual storms, and at the same time track airplanes for air traffic controllers. Like Snow, Kimpel sees scientific and forecasting advantages. "A lot of tornadoes last only a few minutes, but [one of today's radars] takes two sweeps to make sure something is happening. With phased array radar, we hope to see things we haven't seen before."

Filling radar gaps. Neither polarimetric nor phased array radars will be able to fill in the data gaps between today's NWS radars. Radars use microwaves that follow straight lines, they don't follow the earth's curvature. This means that while beams from WSR-88D radars see weather about 140 miles from the antenna, at this distance the lowest beam is about a mile and a half above the ground. In many cases, such as when meteorologists are looking for indications that a tornado might be forming, they need data from closer to the ground.

Four small radars that were tested in Oklahoma between April and June 2007 were the beginning of a planned nationwide network of radars designed to fill the low-level data gaps. They are part of a five-year, $17 million program based at the University of Massachusetts called the Engineering Research Center for Collaborative Adaptive Sensing of the Atmosphere, or CASA. The radars, which use phased array technology, are small enough to be attached to cell phone towers and will have a range of approximately 18 miles. Since they will be an average of only 15 miles apart, the radars will have overlapping coverage.

Weather satellites

While today's technology—rockets, radio communications, devices that capture images—would

Satellites observe the atmosphere, oceans, land, and ice

Satellite images used on television; the Web; and in books, magazines, and newspapers represent only a small share of the atmospheric, oceanic, ice, land, and solar data that satellites collect.

Geostationary satellites

From their orbits 22,238 miles above the equator, the two U.S. GOES spacecraft have the views shown here. While they don't collect useful information around the edge of their views, they track weather systems and collect other data over the western Atlantic, the eastern and central Pacific, South America, Central America, the Caribbean, and all except the northernmost parts of North America. Satellites from other nations cover the rest of the earth except the polar regions, which no geostationary satellite views well.

GOES West   **GOES East**

GOES orbits keep them directly above the dots on the maps.

Satellite orbits

The orbital focus of all artificial Earth satellites is at Earth's center, which means they circle around the earth collecting data far from the nation that launched them. Most nations share environmental satellite data.

Orbit

Gravitational attraction

Focus of orbit

In addition to the visible images of hurricanes and other storms used by the news media, GOES orbiters collect a wide range of infrared measurements that are used to produce information on humidity, temperatures though the atmosphere, and even atmospheric stability.

GOES

Approximately 22,235 miles

Eagle eyes

Each pixel of the GOES infrared sensor is a circle approximately 2.8 miles in diameter on the cloud.

Equivalent of seeing a dot 0.353 millimeters in diameter five to ten times larger than a gnat or flea from 33 feet (10 meters) away.

Energy detected: Equivalent of a 100-watt light bulb 22.6 miles away.

Energy changes detected: Equivalent of differentiating between a 100-watt and a 101.6-watt light bulb 22.6 miles away.

Polar orbiters

In addition to collecting environmental data, NOAA's
Polar Operational Environmental Satellites relay data
from weather stations, buoys, ships, and animals with
tracking beacons. They receive and relay transmissions
from emergency locator transmitters.

Polar orbiters collect visible, infrared,
and microwave measurements used
to produce products such as
temperature profiles though the
atmosphere, humidity measurements,
aerosols and ozone in the air, and
ground conditions including locations
of ice, snow, and vegetation.

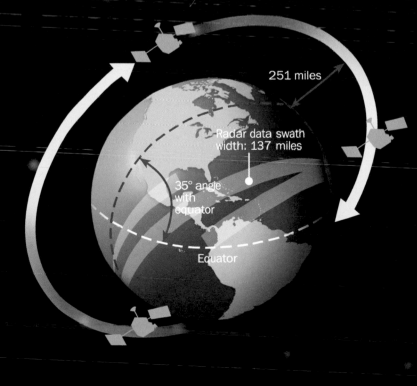

540
miles

Satellite takes
1 hour, 32
minutes to
orbit Earth

Fairbanks,
Alaska

Wallops
Island,
Virginia

Command
and data
acquisition
stations

Night

Each orbit covers
a new swath
because the earth
rotates 25.59
degrees during an
orbit, which is
1,766 miles at
the equator.

Equator

An active radar satellite

The NOAA GOES and polar orbiting environment satellites on
these pages use passive sensing. They detect visible light,
infrared and microwave energy, and solar x-rays the objects
either emit or reflect. Some environmental satellites, such as
NASA's Tropical Rainfall Measuring Mission are active instruments
that send out radar microwaves and detect the energy scattered
back to the satellite.

251 miles

Radar data swath
width: 137 miles

35° angle
with
equator

Equator

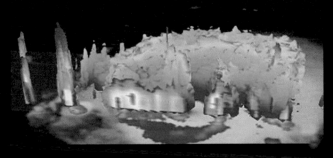

NASA image shows the vertical structure of Hurricane Frances
in 2004 derived from TRMM data

Vendors and government agencies show off equipment for observing the atmosphere and oceans and handling the data at the annual AMS meeting in New Orleans in January 2008.

seem strange to Isaac Newton, he would have no trouble understanding how weather satellites circle the earth. In the seventeenth century he worked out the universal law of gravitation and the basic mathematics today's rocket scientists use to calculate satellite orbits. Newton's key insight was that gravity, the force that makes an apple fall from a tree, also accounts for the moon orbiting the earth and the earth and other planets orbiting the sun.

You sometimes hear that astronauts orbiting the earth have "escaped the pull of gravity." Far from it—the "pull of gravity" is holding their spacecraft in orbit around the earth. A satellite stays in orbit because its inertia—the tendency to continue in a straight line unless a force acts to speed it up, slow it down, or make it follow a curved path—is in balance with gravity. Gravity is the force that makes it follow a curved path, staying close to the earth and not moving farther away. Think of it this way, if you are swinging something attached to a string around your head, the string is acting the same way as gravity acts on a satellite. If the string breaks, the object flies away in a straight line. For the forces of inertia and gravity to be in balance, the satellite must travel at a particular speed in relationship to its altitude above the earth. We'll see why this is important in the section on geostationary satellites, below.

From rockets to satellites. In 1948, Americans launched a captured German World War II rocket at the White Sands Proving Grounds in New Mexico. It carried a camera, which took photographs of clouds over a wide area. Meteorologists who studied the photos saw that they could detect weather patterns. They also realized that tracking the weather from space would require satellites in Earth's orbit, not just rockets that go from one point to another. Both the United States and the Soviet Union planned to launch the first artificial satellite (the moon is Earth's natural satellite) for the International Geophysical Year. On October 4, 1957, however, the Soviet Union launched the world's first artificial satellite, *Sputnik 1*, embarrassing the United States and encouraging a space race to launch bigger and more complex satellites. This benefited meteorology.

The United States launched the world's first weather satellite, the experimental *Tiros 1*, on April 1, 1960, to test sending black-and-white television images back to Earth from a **polar orbit** 450 miles up. Satellites in polar orbit circle the earth in a north-south direction over the Arctic and Antarctic but not directly over the North and South poles. The orbits are tilted. *Tiros 1* operated only 78 days and its black-and-white images were crude by today's standards. But it proved that satellites could be useful. On April 9, 1960, it captured an image of a previously undetected **tropical cyclone** about 800 miles east of Brisbane, Australia.

NASA launched *Tiros 2* on November 23, 1960, with an important addition: a camera that captured infrared images as well as the visual light images that *Tiros 1* had captured. This capability added greatly to the usefulness of satellites and not only because they could now "see" at night. As we saw in Chapter 2, the earth is always emitting infrared energy (day and night), with the frequencies emitted depending on temperature. Today's satellite infrared imagers detect sea-surface temperatures; the temperatures of cloud tops, which correlate well with cloud heights; and even otherwise invisible water vapor in the air.

All of the first weather satellites were polar orbiters, and as such, they had a big disadvantage: they couldn't keep a continuous eye on the weather in any one place, such as an area of strong thunderstorms developing over the plains of Colorado. The science fiction writer Arthur C. Clarke (1917–2008) had proposed a satellite that would solve this problem back in 1945, but he wasn't thinking of weather satellites.

Clark's article in the October 1945 issue of *Wireless World* magazine was entitled "Extra-terrestrial Relays: Can Rocket Stations Give Worldwide Radio Coverage?" Clark was proposing what came to be known as communications satellites. His insight was that if a satellite were put in orbit 22,238 miles above the equator, the speed needed to balance its inertia and the pull of gravity would be 6,802 mph, the same speed as Earth's rotation. If you were standing on the equator, you'd be traveling a little faster than 1,000 mph. Because the satellite's orbit is so much larger than the earth's circumference, the satellite has to travel faster to keep up. Since such a satellite stays over the same place on the equator, it's said to be in "geostationary" orbit. Clark's idea was that three satellites in such orbits could relay radio signals all over the earth. It took almost twenty years, but Clark's dream came true when the U.S. launched the Syncom satellite in time for television coverage of the 1964 Tokyo Olympics.

Two years later, the United States launched a prototype geostationary weather satellite and other experimental satellites, including the 1974 Synchronous Meteorological Satellite with an infrared camera, which proved the value of such satellites and helped lead the United States to launch its first operational (as opposed to experimental) Geostationary Environmental Satellite (GOES) on October 16, 1975. Over the years, technology for satellite sensors and communications has improved, but the basic idea remains the same: satellites that stay over the same location ensure that no big weather events, such as hurricanes, occur without forecasters knowing about it. Meanwhile, polar-orbiting satellites are capturing a more close-up view of the atmosphere from their lower orbits and supplying data to computer models.

Summary and looking ahead

As we've seen in this chapter, huge amounts of weather data are collected around the globe each hour of every day even in remote areas such as on the Antarctic Ice Sheet. In Chapter 7, we will see how global weather data are entered into computers that produce forecasts. But entering the large volume of data into computers in ways that enable them to provide useful results is a science unto itself. It's not easy.

Different kinds of data have their own strengths and weaknesses. As we've seen, a rain gauge gives a reasonably accurate report of how much rain falls on a particular location, while weather radar gives an overall picture of an area's rain, but one that is not as accurate. Satellites see big weather pictures but can't directly measure a location's wind or atmospheric pressure. Computers need numbers; they cannot read satellite or radar images as meteorologists can. Over the years, scientists have developed ways for computers to use satellite data and are now working on making radar data more useful.

"I like to use the analogy," says Louis W. Uccellini, director of the NWS National Centers for Environmental Prediction, "that you go to a supermarket to buy food, you go a restaurant to buy a meal. The chef has put in effort to take food and make it a meal." The men and women who design and operate **data assimilation** systems (that is, systems that combine data from multiple instrument types) are the chefs that prepare the food (data) as easily digestible meals (forecasts).

With what we've learned in this chapter about how weather is observed and in previous chapters about the basic science of weather, we're ready to see how weather is forecast in Chapter 7.

Weather forecasts from start to finish

All weather forecasts begin with collecting data about the current state of the atmosphere, the oceans, and the earth's surface and sending the data to forecasters. While meteorologists can produce short-term, localized predictions using local data, most of today's forecasts seen on television or the Web as well as the specialized forecasts, such as for aviation, are based on global data and forecast products (maps and text) from the U.S. NWS Centers for Environmental Prediction based in the Maryland suburbs of Washington, D.C., or from similar centers in other nations.

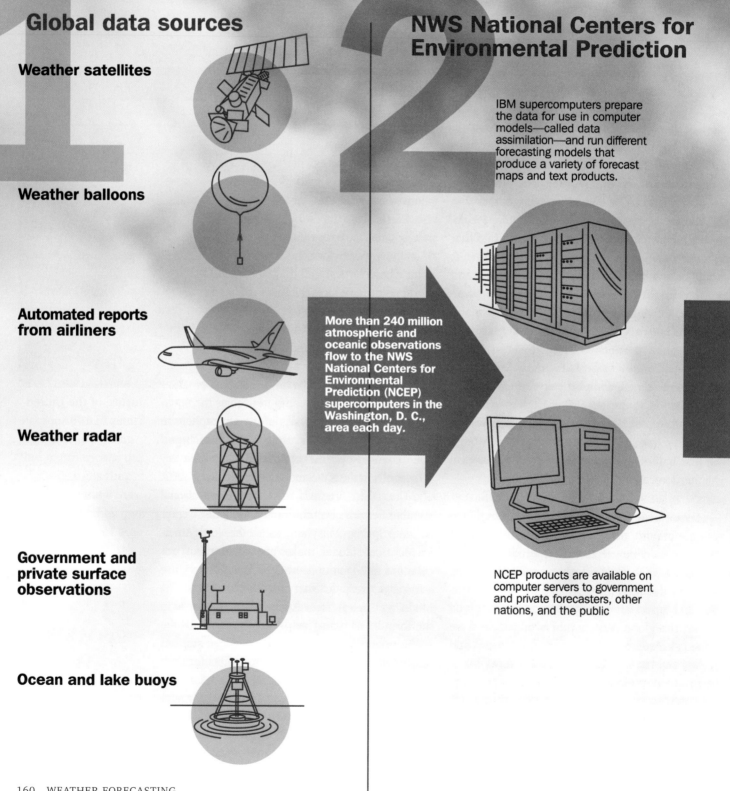

Global data sources

Weather satellites

Weather balloons

Automated reports from airliners

Weather radar

Government and private surface observations

Ocean and lake buoys

More than 240 million atmospheric and oceanic observations flow to the NWS National Centers for Environmental Prediction (NCEP) supercomputers in the Washington, D. C., area each day.

NWS National Centers for Environmental Prediction

IBM supercomputers prepare the data for use in computer models—called data assimilation—and run different forecasting models that produce a variety of forecast maps and text products.

NCEP products are available on computer servers to government and private forecasters, other nations, and the public

The forecasting enterprise

Meteorologists generally refer to National Centers for Environmental Prediction products as "guidance" because they aren't always the final word for localized or specialized predictions. The organizations in this section are some of the many that make use of NCEP products. Some of these organizations also run their own forecasting models or use products from other centers such as the European Centre for Medium-Range Weather Forecasts in Reading, England, or from universities and research centers.

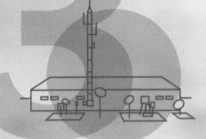

122 NWS regional offices

Issue weather warnings such as for tornadoes. Produce public forecasts for their areas.

U.S. military forecasters

Produce forecasts for land, sea, and air operations.

NWS Storm Prediction Center

Norman, Oklahoma

Issues tornado and severe thunderstorm watches. Produces national fire weather products.

Private forecasting companies

Provide specialized forecasts for a variety of businesses including air lines, shipping companies, power companies, and the news media.

NWS National Hurricane Center

Miami, Florida

Forecasts tropical storms and hurricanes in the Atlantic, Caribbean, Gulf of Mexico, and Pacific east of 140° W longitude.

Television meteorologists

Use NCEP and local NWS office information and sometimes images or data from private firms to produce forecasts.

21 NWS aviation center weather service units at air traffic control centers.

Provide air traffic controllers with weather information.

World Wide Web and newspapers

Private companies or television meteorologists produce forecasts for newspapers and online Web sites. NWS products are also available on the Web.

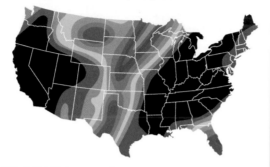
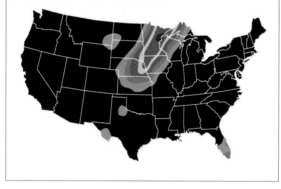
weather elements across the United States, such as the amount of liquid precipitation—rain and melted ice or snow—during six-hour periods ending at specific times.

Analyzing results. Steve Lord, director of the NWS Environmental Modeling Center in Camp Springs, Maryland, says good forecasts require huge amounts of data, efficient ways of feeding the data to the computers, the best-available computers, and a post-processing system for keeping tack of errors. "Ninety-nine times out of one hundred we do quite well," he says. But, Lord adds, when a forecast goes wrong the complexity means "it's very difficult to go back and figure out exactly what went wrong." This is because today's weather forecasting systems "are so complex that they rival the atmosphere in their complexity. The reason we are so good at forecasting is that we are getting the complexity of the atmosphere pretty well."

This complexity is one reason why computer modeling is concentrated at places such as the center in Camp Springs. When a problem crops up, experts in different disciplines such as physics, mathematics, oceanography, meteorology, and others work together to solve it. Often, Lord says, the best bet isn't to chase the reason for a particular error, but to improve the overall system with the goal of increasing the odds of producing better forecasts.

Tightening grids. While grids of forecasting models are steadily becoming tighter, grid points are still miles apart. However, some of the important processes, such as condensation of water vapor, happen on the molecular level. While scientists model molecular-level processes for various kinds of research, even today's super-fast computers wouldn't be able to handle all of the calculations needed to get the details exactly right for each rain shower.

Also, even if complete weather observations could be taken at every point of the model's grid, which is impossible with any current technology, important things could be going on between grid points. For example, a small shower that will grow into a thunderstorm could be occurring between two observation points without being seen.

For the same reason, the large-scale models can't predict exactly where thunderstorms will

puters. Computer-generated MOS forecasts use vast amounts of stored, historical data to correlate values of dozens of past weather parameters (such as temperature, air pressure, and wind at different altitudes) with the subsequent values of those parameters. MOS forecasts are a much more sophisticated use of climatology than merely forecasting that tomorrow's high and low temperatures will be the 30-year averages for the day. MOS correlation strategies are constantly updated and refined to improve future forecasts of the parameters.

MOS forecasts include text tables listing the expected temperatures, humidity, winds, amount of cloud cover, the probability and type of precipitation, and more at specific locations around the country each three hours for the following 72 hours. MOS output also includes maps showing

occur, only the general regions and times when upper-air conditions will be ripe for thunderstorms and how powerful any thunderstorms that do form could be. This is why the weather, and in particular small but dangerous phenomena such as thunderstorms and tornadoes, can surprise forecasters.

Forecast modelers cope with factors that would slip between the grid points with what is known as **parameterization**. Thunderstorms provide a good example. If a model shows that conditions will be good for thunderstorms to form at a certain time and place, the model can, in effect, "create" the thunderstorms. It then feeds data back into the grid points surrounding a "thunderstorm" describing the heating, added humidity, air movements, and precipitation associated with it.

In 2004 the NWS modeling center began using the Weather Research and Forecasting (WRF) Model, which is being used by weather services around the world as well as by researchers. One of its advantages is that it can be operated on a wide range of scales, including those that focus on a relatively small area, such as a couple of Great Plains states. The closer grid spacing and other advantages of this model mean it can be used to forecast relatively small events, such as lines or clusters of thunderstorms. In fact, it can include programs that handle the cloud microphysics calculations of thunderstorms.

Forecasting limits

Even with the advances in data gathering, ever-faster computers, and continual improvements in models, anyone trying to forecast the weather faces some formidable obstacles, as illustrated by the differences between forecasting an eclipse of the sun and the weather on the day of the eclipse. Astronomers can say with certainty that a spectacular, total eclipse of the sun will race across the United States from northern Oregon to the Atlantic Coast near Savannah, Georgia, on August 21, 2017.

Meteorologists don't expect to be able to predict with much certainty a week before August 21, 2017, whether or not clouds will block the view of the eclipse at particular locations along its path.

Steve Lord says he expects that by 2017 six- or seven-day forecasts will be as good as today's five-day forecasts. But weather forecasts certainly won't be good enough even then to help someone plan a

> **Eclipse predictions**
> To make their predictions of eclipses, astronomers calculate the positions of the moon and the sun relative to the earth as the gravitational force of each is acting on the other two. Such a "three-body" problem requires complex calculations, but as the long tradition of eclipse prediction shows, they are done with great precision and confidence. Compared with the weather, the major bodies involved—the earth and the moon—are so large and have so much momentum that an extremely large force would be needed to change the orbit of either.
>
> Weather events, on the other hand, operate on scales from the molecular level (latent heat released when water vapor condenses) to the global (the Hadley cell). This is one of the reasons why, as we see below, chaos plays such a large role in weather but not in eclipse predictions.

tour several months ahead of time and select the best location for a group to watch the eclipse (where clear skies will allow a good view). The only way to make such decisions even in 2017 will be by consulting climate data, which shows that places between the Cascade Range in Oregon and the Rocky Mountains have more sunny August days than locations along the eclipse path east of the Rockies. In fact, Lord expects that in 2017 clouds will continue to be one of the most difficult phenomena to forecast.

Chaos complications. The atmosphere is a complicated system with several variables affecting each other in different ways. One effect of the atmosphere's chaos is that predictions of its future state—the temperature, pressure, humidity, and so forth—at a particular time and place have "a sensitive dependence on initial conditions," in the words of Edward Lorenz (1917–2008), a theoretical meteorologist at the Massachusetts Institute of Technology. Lorenz discovered that the atmosphere's chaos puts a theoretical limit on the accuracy of weather forecasts. Chaos means that tiny differences now can lead to big differences in the future.

In 1961 Lorenz was conducting experiments with an early computer model that reproduced some aspects of the weather. One day he decided to repeat some computations to examine the model in greater detail. "I stopped the computer, typed in a line of numbers that it had printed out earlier, and set it running again," he writes in his book *The Essence of Chaos*. Lorenz says he "went down the

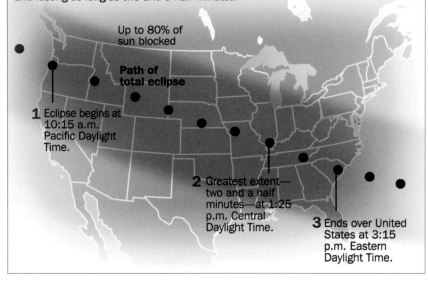

Eclipse forecast for 2017

The Monday, August 21, 2017, eclipse of the sun will sweep across the United States with the moon completely blocking the sun in a path approximately 70 miles wide and lasting as long as two and a half minutes.

Up to 80% of sun blocked

Path of total eclipse

1 Eclipse begins at 10:15 a.m. Pacific Daylight Time.

2 Greatest extent—two and a half minutes—at 1:25 p.m. Central Daylight Time.

3 Ends over United States at 3:15 p.m. Eastern Daylight Time.

hall for a cup of coffee and returned after about an hour" to discover that the "numbers being printed were nothing like the old ones." Before calling someone to service the computer, he examined the numbers and found that those calculated after he had restarted the computer at first repeated the old numbers but then began to differ by large amounts over time.

Lorenz realized that the numbers he had typed in were not exactly the same values as the old ones but were rounded off. That is, the computer calculated to six places but printed out to only three places, and he had used these values when he typed in the values before restarting the computer. The small differences in values made little difference at first, but the model's results diverged increasingly with time. Small changes in the information fed into the model led to big changes in its output.

Since the 1960s, chaos theory has become an important part of many sciences, not just meteorology. For example, the American Institute of Physics publishes *Chaos*, a quarterly journal. Titles of articles in 2005 and 2006 issues of the journal included terms such as epileptic seizure prediction, music recommendation networks, computer performance, turbulence, orbiting planetary satellites, and detecting damage in a structure such as a bridge.

Popular culture has picked up one of terms that

Lorenz used: "the butterfly effect." It comes from the program for a 1972 American Association for the Advancement of Science meeting that gave the title of Lorenz's talk as: "Predictability: Does the Flap of a Butterfly's Wings in Brazil Set Off a Tornado in Texas." In *The Essence of Chaos*, Lorenz argues: "If the flap of a butterfly's wings can be instrumental in generating a tornado, it can equally well be instrumental in preventing a tornado."

While Lorenz and others began the work of making chaos a formal mathematical study, meteorologists had been aware of it for decades in a much more informal way. In his book, Lorenz mentions that his sister gave him a copy of George Stewart's 1941 novel *Storm* when she heard he was going to study meteorology. Lorenz says that in the novel, "a meteorologist recalls his professor's remark that a man sneezing in China may set people to shoveling snow in New York. Stewart's professor was simply echoing what some real-world meteorologists had been saying for many years, sometimes facetiously, sometimes seriously."

Ensemble forecasting. One result of the weather's inherent chaos is that, as Steve Lord of the Environmental Modeling Center puts it, "Sometimes we get the weather just right. Sometimes there's almost nothing we can do to get it right." Since 1992, forecasters have been using the growing amount of available computer power to find some order in the atmosphere's chaos, or to at least characterize its chaos by using the technique of **ensemble forecasting**.

To produce an ensemble forecast, meteorologists run a computer model perhaps a hundred times, making slight changes in the initial data for each run. Forecasters then apply statistical techniques to the results to produce measures of how reliable the forecasts are likely to be. While this can become quite complicated, one kind of ensemble forecast, appropriately called **spaghetti plots**, as shown in the graphic on the next page, gives you a feel for how ensemble forecasting works.

The basic idea of comparing forecasts is, of course, somewhat like the idea of obtaining a second opinion about an illness, except weather modelers can obtain hundreds of second opinions and use their computers to sort them out. Joel K. Sivillo and Jon E. Ahlquist of Florida State University and

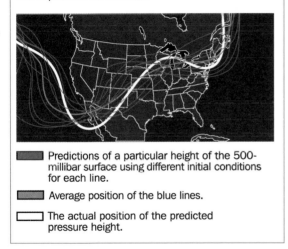
Zoltan Toth of the Environmental Modeling Center argue in a 1997 article in the journal *Weather and Forecasting* that the forecast for the June 6, 1944, D-Day invasion of France was "an earlier example of ensemble forecasting…which considered predictions from three independent forecasting teams using different techniques. As the D-Day forecast example shows, ensemble forecasting does not have to involve computers. If two or more human forecasters make separate forecasts and then [compare] them, they are working with an ensemble of forecasts."

One of the major uses of ensembles is to give forecasters an idea of how much confidence they should have in a particular forecast. If a spaghetti plot looks like someone carefully placed the strands on a plate, forecasters can be more confident in their predictions.

Meteorologists are talking more now about finding ways to indicate their confidence in forecasts. The Capital Weather Gang blog on the *Washington Post* Web site offers an example of how this can be done. Each forecast is labeled with how much confidence the forecaster has in it. These are:

High: Bank on it.

Medium-High: Overall forecast is sound but minor variations are possible.

Medium: We think we're on the right track, but the forecast details are still taking shape.

Low-Medium: This is our best guess but don't hold us to it.

Low: Crapshoot.

Jason Samenow, the blog's site manager and chief meteorologist says the "confidence interval is subjective—based on how comfortable we feel making the forecast." The blog's meteorologists consider forecast ensembles and differences among the major operational models, both from the NWS and the Centre for Medium-Range Weather Forecasting in Reading, England. "If there is a large spread in the ensembles or a lack of agreement in the various models, forecast confidence is reduced. And, of course, typically our forecast confidence bars will show decreasing confidence for the longer range forecasts. But depending on the weather pattern, short term confidence isn't necessarily always high."

Delivering forecasts

Computers are not only revolutionizing how weather forecasts are made but also how meteorologists deliver them. Lynn Maximuk, who became director of the NWS Central Region in 2006, says that before computers became so ubiquitous, the Weather Service was "limited to products and the forecasters were challenged to put everything in their heads into two or three paragraphs." Now, however, he says, "For the National Weather Service, the primary product is now information, not text messages. Since the nineteenth century,

Sverre Petterssen, one of the meteorologists who helped make the D-Day forecast tells his story in *Weathering the Storm: Sverre Petterssen, the D-Day Forecast, and the Rise of Modern Meteorology*, edited by James Rodger Fleming, published by the AMS in 2001.

Weather forecasters call the results of their work "products." These include warnings of dangerous weather, brief predictions for the public, tables of forecast data, complex charts of expected conditions, and discussions of the reasoning behind meteorologists' forecasts.

A journalist and educator

The millions of people who watch Tom Skilling's weather presentations on WGN-TV, read the *Chicago Tribune* weather page with his forecasts and answers to questions, or click to his blog know he's more than a television personality.

Tom Skilling says, "I've always worked twelve, fourteen hours a day. It's all consuming; it's a labor of love."

The millions who watch him on WGN America around Chicago and elsewhere via cable and satellite "certainly aren't tuning in for my pretty face," Skilling says. "I view myself as a journalist first and foremost, and my beat is the atmosphere."

He says television meteorologists are educators with the duty of alerting people to forecasts and warnings. As an educator, Skilling produces documentaries on topics as diverse as tornadoes, lightning, weather forecasting, Alaskan meteorology, hurricanes, and tornadoes in addition to his broadcasts, the newspaper page, and his blog.

In 1991, his first documentary, *It Sounded Like a Freight Train*, about tornadoes, earned Super-Bowl-like ratings and received several awards, encouraging WGN to have him produce more documentaries.

Skilling first became a broadcast meteorologist for an Aurora, Illinois, radio station when he was 14. He moved to television in Aurora when he was in high school. Skilling continued as a part-time television meteorologist at the University of Wisconsin–Madison in the early 1970s. At the time the university was doing leading-edge weather satellite and weather-data displays research, including satellite loops. "Wouldn't this be wonderful if you could bring (these displays) to television to show people what's happening," Skilling thought.

Before Skilling earned his degree, a Jacksonville, Florida, station offered him a job. "Do I stay here and finish my degree, or do I go down there and fly with the hurricane hunters, learn to shoot film, and get some tropical meteorology experience?" He decided to go, but first passed the day-long written test the AMS then required for its broadcaster's seal of approval. He also worked in Milwaukee, Wisconsin, before going to WGN in 1978. He has been organizing free public tornado and severe weather seminars each spring since 1981 at the Fermilab National Accelerator complex in Batavia, Illinois. "One of the highlights of my career."

"I'm still self-conscious about not having a degree…I don't regret having done it that way, but I had to work harder to prove myself to meteorologists. The decision I made at the time was not a bad one. I had chances for experiences people couldn't have today. Without a degree, they wouldn't get in the door." He has proved himself. For instance, he's a Fellow of the AMS, which recognizes "outstanding contributions" over a "substantial period of years."

When Skilling started in Aurora, his forecasts were based on 12-hour-old weather maps the Chicago Weather Bureau office mailed to him via special delivery. At the Aurora television station he learned to read the weather from the Weather Bureau's cryptic, coded teletype messages. The leading-edge weather displays he saw at the University of Wisconsin in the 1970s now seem antique and he's amazed at the advances since then. "Look at the revolution we've seen in our ability to sense the atmosphere. I think there has been more progress in meteorology in the last forty years than in all of the time before going back to Aristotle. Today, we have a fighting chance of understanding how nature [puts] together the weather patterns we see."

weather offices have been issuing forecasts at regular times of the day and alerts and warnings whenever dangerous weather threatened. These are still being done and are likely to continue, but weather consumers are seeing an increasing number of choices."

Now, however, Maximuk says, both the NWS and private forecasting firms are finding new ways to use the NWS's digital database of forecasts and other data. For example, a private forecasting company working for a county highway department could feed forecasts from the Weather Service's databases, such as when precipitation should begin and end, to the company's own decision-making computer models that help highway managers decide the most efficient way to respond to an approaching snow storm.

Before becoming Central Region director, Maximuk was the meteorologist in charge of the Pleasant Hill, Missouri, NWS office. That office, like many others, is experimenting with ways to help users make weather-related decisions. "We're allowing our customers and partners to take information and use it and format it in ways that meet their needs," Maximuk says. "They are no longer [dependent upon] the Weather Service to provide forecasts."

Highway weather

Weather that disrupts highway travel affects millions of people every year but usually attracts little public attention except after a (fortunately rare) multiple-vehicle pileup on a fog-bound interstate highway.

Road weather is relevant to much more than such disasters. For example, when NOAA and the Federal Highway Administration signed a memorandum of understanding in July 2005 to work together on highway weather issues, they said the goal is to "help mitigate the 7,000 deaths, more than 600,000 injuries, 1.4 million crashes, and $42 billion in economic losses that occur each year due to adverse weather."

In its 2004 report entitled *Where the Weather Meets the Road: A Research Agenda for Improving Road Weather Services,* the National Research Council offers an imaginary scenario for Lou, the driver of semitrailer tractor heading north on Interstate 35 across Iowa toward Duluth, Minnesota, a few

years from now. It describes how systems then could handle a major winter storm that threatens to block his route. In the scenario, Lou and his dispatcher use the truck's "always-on" communication system to decide whether he should continue driving, stop and wait until the storm passes, or detour to the east.

The dispatcher uses a continuously updated forecast tailored for the trucking company to advise Lou to continue north on I-35 since the storm is forecast to stay west of his route. As often happens, however, the weather changes after a couple of hours and the forecast now says the storm could cause patchy snow and ice on the road. In fact, several vehicles about a mile ahead of Lou detect highway ice and report this via the highway's "intelligent" network.

The main communication computer on Lou's truck picks up these reports, alerts him with a yellow light, and automatically reconfigures his truck with better traction. The truck's electronic stability control system does this by adjusting the power and breaking forces applied to each wheel to control skids. (Such systems are on many of today's cars and trucks.) The dispatcher keeps Lou updated about plowing and chemical anti-ice treatments of the highway, automated spray of chemicals to anti-ice bridges, and traffic and visibility, as well as the progress of the storm as Lou continues without making unnecessary and costly stops.

At first glance, this vision might seem as unlikely as some of the projections in the "Miracles You'll See In The Next Fifty Years" article in the February 1950 edition of *Popular Mechanics* magazine, such as: "Storms are more or less under control...With storms diverted where they do no harm, aerial travel is never interrupted."

In contrast to 1950s dreams of the future, by 2004 most of the technology needed for the scenario in the National Research Council report was in operation; no scientific breakthroughs or new technology are needed to turn the scenario into reality. Automobile manufacturers are using or experimenting with the devices needed for vehicles to play their role in the scenario. Highway departments around the United States and in other nations are using many of the required elements, and researches are refining various parts of the needed systems.

Storms that are relatively small,

but which are sometimes fierce,

destructive, and deadly

CHAPTER 8

A little before 2:30 the afternoon of July 13, 2004, Craig Joraanstad stepped out of his office at the Parsons Company, Inc., factory near the small town of Roanoke in central Illinois for a clearer view of the menacing clouds outside of his office window.

For the first time in his life, Joraanstad saw rotation in a cloud—a "very large, dark mass of clouds rotating out to the northwest." When he saw a thin cloud drop from the mass of clouds and kick up dirt when it hit the ground, he said to himself, "Holy cow!"

If Joraanstad had been at almost any other factory, or a school, shopping center, church, or hospital with 150 people inside, we can imagine him running inside shouting, "Tornado! Tornado!" Some of those inside would rush to windows to look; others would squeeze under desks, crowd into windowless restrooms and closets, or stand paralyzed. In the next ten minutes the **tornado** could kill or seriously injure many of them.

That didn't happen in Roanoke on July 13, 2004, because of decisions made by Bob and Terri Parsons, owners of the factory more than thirty years before, and that afternoon's actions by Parsons employees and NWS forecasters.

When Joraanstad first noticed the dark clouds from inside his office, Tony Hall, a forecaster at the NWS Lincoln, Illinois, office, forty-six miles to the south, was preparing a **severe thunderstorm warning** based on weather radar images of the same thunderstorm.

At 2:29 p.m., the weather radio in the Parsons accounting department sounded with Hall's severe thunderstorm warning. As called for in the company's long-standing plan, Laura Marchand in the accounting department notified the company's emergency coordinator, Dale Eastman, who alerted weather spotters. Only two minutes later the spotters saw the rotating cloud about five miles to the northwest and told Eastman. He asked Patricia Canon in the accounting department to announce over the factory's loud speakers that everyone should go to the factory's shelters.

As Joraanstad went back inside, he heard Canon making the announcement. During her announcement, the weather radio sounded with the **tornado warning** that Hall had started preparing after issuing the severe thunderstorm warning.

Previous pages: Lightning hits the side of the Washington Monument in Washington, D.C., on July 1, 2005, showing that it doesn't always strike the highest point in an area. Lightning is always a thunderstorm danger. Some also produce tornadoes, other dangerous winds, large hail, or flooding rain.

Only three minutes after the announcement, company supervisors confirmed that everyone at the factory, 140 employees and 10 visitors, was in the shelters.

As the tornado crossed Illinois Route 117 in front of the factory, it strengthened from an F3 tornado with winds between 158 and 209 mph to a quarter-mile-wide F4 with winds estimated as powerful as 240 mph. It smashed into the factory at 2:41 p.m., seven minutes after the take-shelter announcement and the tornado warning.

The lights went out as the tornado hit and the sound was "like when you take the cap off an empty water bottle and crush it, only louder," Joraanstad said. The tornado was ripping all of the sheet metal from the roof. People were praying.

Some in the shelters held their hands over their ears to block out the roar and banging. The sudden decrease in air pressure hurt their ears and some had earaches for days afterward. People in one shelter were in a circle holding hands. As the noises grew louder, some feared they were about to die.

When the clatter of large and small objects hitting the shelter ended, "it was completely dark; we didn't know what to do," Joraanstad said. "We heard gas escaping from broken high-pressure lines outside the shelter." Those inside had seen electrical wires outside the door and feared sparks from them would ignite the gas. "Then someone realized we didn't have any power—there wouldn't be an explosion." They picked their way through the wreckage to the assigned meeting area in the parking lot.

"We were all in shock," Joraanstad recalled. Someone shouted, "Where are the cars?" The tornado had thrown all of the cars from the parking lot into the wrecked building. Without contact with the other shelters, those in Joraanstad's group didn't know whether anyone else survived until they saw "people coming single file through the wreckage from the other shelters, looking like little rows of ants." Not only did all 150 people in the factory live, none were injured.

As the shock of the tornado wore off and the fact that no one was dead or injured sunk in, those gathered in the parking lot began to wonder: Will the plant be rebuilt? Joraanstad, who is the Parsons human resource director, said, "My fear was it's taken me so many years to hire these great, talented people; we're going to lose them all. My

The Enhanced Fujita Scale	Tornado EF rating	Wind gusts in mph
The Fujita Scale is based on a tornado's damage and is determined only after the tornado passes when NWS or other experts study the damage. The NWS adopted the Enhanced Fujita (EF) Scale on February 1, 2007. The wind speeds listed in the table are wind gusts that last at least three seconds.	0	45–78
	1	79–117
	2	118–161
	3	162–209
	4	210–261
	5	262–317

other fear, personally, is what am I going to do? I'm nearly fifty years old. What am I going to do?"

When the tornado hit, Bob Parsons was in Fort Worth, Texas. Kevin Trantina, the company's chief operations officer, called on his cell phone from the parking lot.

"We've been hit head-on," Parsons heard. "We've been hit head-on, Bob, by a tornado. The plant is totally destroyed."

"What about all the people?" asked Parsons.

"I think everyone is accounted for," Trantina answered.

Parsons' hobby is training and showing cutting horses, which are used to "cut" or separate individual cows from a herd. He was in Fort Worth competing in the National Cutting Horse Derby. "I was in the finals, which are hard to get into. I walked away and didn't even compete. It wouldn't have been right." He chartered a jet and about three and a half hours later was flying over the ruins of the business he started building in the early 1970s. "I couldn't believe what I was looking at."

Three days later, at an all-employee meeting at a local grade school, Parsons told everyone that the factory, which manufactures parts for Caterpillar, Inc., and other companies, would be rebuilt and everyone would continue receiving his or her regular salary or wages. In fact, Joraanstad said, hourly workers were paid not for their regular forty hours a week, but the weekly average they had earned with overtime pay the previous year.

While the factory was being rebuilt, some employees worked in leased locations producing parts for Caterpillar, Inc., and other customers; some cleaned up debris and assisted in rebuilding. Others were sent into surrounding communities for projects such as making repairs or painting at nursing homes and schools.

A severe thunderstorm is one that has winds of 58 mph or greater at the ground, has hail at least three-quarters of an inch in diameter, or produces a tornado.

that can cause flash floods, hail, and, of course, tornadoes.

Fortunately, violent thunderstorms are rare, and only a few produce tornadoes, strong straight-line winds, or hailstones large enough to kill someone. Humans rarely encountered the dangers (other than lightning) of garden-variety thunderstorms until scientists, engineers, and aviators developed **instrument-flying** techniques in the 1930s that allowed pilots to safely penetrate clouds almost all of the time. Instrument flying opened the door to safe scheduled airline service and "all-weather" military flying during World War II, but it exposed pilots and their passengers to dangers unknown on the ground, such as severe **turbulence**.

The July 28, 1943, crash of an American Airlines DC-3 in a thunderstorm near Bowling Green, Kentucky, which killed all twenty-two aboard, was one of several crashes before and during World War II that focused attention on the dangers thunderstorms pose to airplanes; dangers that may not reach people on the ground. For instance, the Civil Aeronautics Board, a forerunner of today's U.S. Federal Aviation Administration, said in its report on the Bowling Green crash that the DC-3 was about 1,300 feet above the ground while attempting to land when a violent thunderstorm **downdraft** pushed it down abruptly, causing it to crash into trees.

Thunderstorm science. The Bowling Green crash, other airline crashes in the 1930s and 1940s, and military crashes during World War II prompted Congress to fund a major thunderstorm study. In 1946, the U.S. Weather Bureau, with major assistance from the military and the National Advisory Committee for Aeronautics (NASA's predecessor),

When firefighters and police reached the top of a nearby hill and saw this wreckage of the Parsons Company factory shortly after the tornado hit, they thought they'd find scores of injured and dead victims. Not one of the 150 people in the factory was injured.

Less than a year after the tornado hit, a new Parsons Company factory was operating on the site where the tornado had destroyed the old plant. Keeping everyone on the payroll and rebuilding the factory "was the thing to do," Bob Parsons says. Parsons Company workers "do custom, specialized work and it's highly skilled." For such a business to be successful, "you've got to have good people and take good care of them."

Thunderstorms

Strong tornadoes, like the one that destroyed the Parsons Company factory, come from violent thunderstorms known as **supercells**. While a supercell can be especially deadly, all thunderstorms are dangerous.

It's safe to assume lightning from thunderstorms started some of the first fires humans used for warmth and cooking. We can also be sure that humans learned early on that the lightning, which brought them fire, could also kill them.

In addition to lightning, thunderstorms can kill or injure people and animals and damage crops and property with their fierce winds, downpours

The life cycle of thunderstorms

One of the major discoveries by scientists who conducted the 1946–1947 Thunderstorm Project was that thunderstorms go through the life cycle shown here. While many thunderstorms go through this cycle as individual, isolated cells, more become part of multicell clusters, or even grow into supercells.

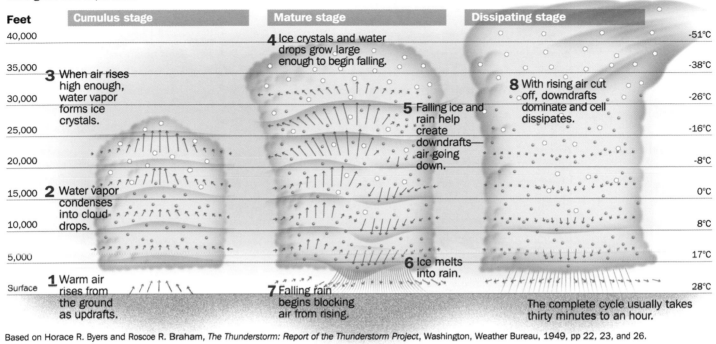

Based on Horace R. Byers and Roscoe R. Braham, *The Thunderstorm: Report of the Thunderstorm Project*, Washington, Weather Bureau, 1949, pp 22, 23, and 26.

organized the Thunderstorm Project, which attracted some of the nation's best university researchers.

Horace R. Byers of the University of Chicago directed the project, which began during the summer of 1946 a few miles south of Orlando, Florida. "Basically, the observation program was designed to obtain a complete description of the thunderstorm and measure its intensity," Byers wrote in the September 23, 1949, issue of *Science* magazine. In addition to seeking information that would aid military and civilian aviation, the researchers also set out to answer basic scientific questions about thunderstorms.

In 1946, the project included fifty-four stations approximately a mile apart in a seven-by-thirteen-mile area near where today's Disney World is located south of Orlando, Florida. These stations automatically recorded weather data. The project also used weather balloons and weather radars. For the first time, airplanes equipped to measure factors such as temperature and humidity were deliberately flown into thunderstorms as radar probed the storms and weather observations were recorded on the ground below. The U.S. Army Air Forces sup-

plied sturdy Northrop P-61 "Black Widow" night fighter airplanes, "flown by some of the Air Force's most expert instrument pilots, who volunteered for the task," Byers wrote. In 1946 the airplanes flew from the Pinecastle Army Air Base south of Orlando, which is now Orlando International Airport. For the summer of 1947, the Project moved to southern Ohio to operate from the New Castle Army Air Base near Wilmington.

The basic research plan required that at least five airplanes fly through each thunderstorm at altitudes 5,000 feet apart from 5,000 to 25,000 feet above sea level. "An effort was made to perform these flights through the most vigorous thunder-

Project pilots and airplanes

Thunderstorm Project pilots were in the U.S. Army Air Forces when the project started. In September 1947, this part of the Army became the separate U.S. Air Force, the name Byers used in his Science article.

Their Northrop P-61s were built as radar-equipped night fighters. One of the Thunderstorm Project's P-61s is on display at the Smithsonian Air and Space Museum's Steven F. Udvar-Hazy Center, adjacent to Washington's Dulles International Airport.

while the idea is simple enough, a lightning rod needs to be installed correctly and then maintained properly to avoid problems, such as electricity arcing across a gap in the conductor and starting a fire.

Anatomy of lightning. A lightning stroke begins with a **stepped leader** working down from the cloud toward the ground. As the stepped leader nears the ground, one of its many branches establishes a connection with the ground, usually through a tall conducting object in the immediate area. The stroke's full blast of energy, or discharge, then follows this initial path to the ground. As Martin Uman says in his book *All About Lightning*, "a complete lighting discharge is called a **flash**... [and] typically lasts a few tenths of a second." Each such brief flash typically has three or four **strokes**, but some have as many as thirty or forty strokes with each stroke between the cloud and the ground or the ground and the cloud carrying electricity. Most of the time the lightning we see is the bright **return stroke**, which travels from the ground to the cloud immediately after the dim stepped leader reaches the ground.

If lightning's electrical current passes through the wood of a building, the wood's resistance to electrical current can heat it enough to burst into flame. Lightning rods protect buildings by giving lightning discharges a good conducting path to the ground.

Lightning often hits living trees without setting them on fire because rainwater on the tree or the water in its inner bark gives electricity a low-resistance path to the ground, but this often leaves scars running down the tree's bark. At times, however, the discharge passing through a live tree can flash-heat the water inside the tree to steam, blowing bark and chunks of wood from the tree. Lightning that hits dried-out, dead trees or brush starts wildfires. Lightning scientists talk of "hot" and "cold" lightning even though, as Uman points out in *All About Lightning*, any lightning stroke has a temperature of 15,000 to 60,000 degrees Fahrenheit. The difference between hot and cold lightning is how long the stroke lasts, with a cold lightning stroke lasting only thousandths of a second. A hot stroke lasts tenths of a second (much longer), which can be long enough to ignite wood. This doesn't mean a cold stroke won't cause damage. Such strokes can instantly turn water into steam or even vaporize some solids, such as plastic, quickly enough to blow the material apart, Uman says.

The growth of lightning science. While electrical science and engineering made great strides during the nineteenth century, little was added to what Franklin and his contemporaries learned about lightning until the twentieth-century development of high-speed cameras and instruments to measure the electric and magnetic fields lightning produces. The proliferation of large power grids and their increasing vulnerability to lightning strikes has encouraged more intensive research into the phenomenon.

The March 17, 1934, issue of *Science News* describes one of the big advances, B. F. J. Schonland of Cape Town University and H. Collins of the Victoria Falls and Transvaal Power Company (in South Africa) found some answers to the question, "How fast does a thunderbolt travel?"

They used a camera developed in 1926 by the British physicist C.V. Boys to capture the entire stepped leader–return stroke cycle. The speeds of the leaders they measured ranged from 810 to 19,900 miles per second, and the leaders averaged 5.15 miles long. The lightning return strokes photographed ranged from 1.6 to 4.7 miles long with the quickest taking 69 microseconds to travel 3.5 miles.

Creating lightning. Capturing images or electric and magnetic field measurements very near lightning strokes is a matter of luck as well as science. The odds of lightning hitting any particular place where scientists are waiting to measure it are

For each lightning flash that hits the ground, three to five stay within a cloud or flash from a cloud to another cloud or to clear sky.

low. For example, the University of Florida's International Center for Lightning Research and Testing on the Camp Blanding Army National Guard Base near Starke, Florida, is in one of the most lightning-prone parts of the United States. Yet natural lighting hits the one-hundred-acre site only a half dozen times a year.

To ensure lightning hits targets, such as power lines or a small house—both built for lightning tests—and near enough to the instruments and imagers for good readings, Center researchers launch three-foot-long rockets toward clouds overhead that are ready to produce lightning. Each rocket trails up to 2,000 feet of Kevlar-coated wire to trigger lightning by supplying an electrical path to the ground, much like that of a stepped leader. Center researchers launch rockets seventy-five to eighty times a year with only half of them triggering lightning.

Lightning's mysteries. Anyone who is fascinated by lightning and has an affinity for mathematics and science can expect to have much to explore as an undergraduate, graduate student, and researcher. Scientists aren't likely to answer all of the big questions about lightning before today's school children begin their careers as tomorrow's researchers.

Descriptions of how clouds become electrified and how lightning occurs are very generalized, based on what scientists now know. When you begin looking more deeply into lightning, however, you find that scientists have more questions than answers.

For instance, Uman says, "Nobody really knows how lightning gets started in clouds. Electric fields big enough to start lightning like those that make long sparks in laboratories have never been observed in a cloud. On the other hand, it's hard to put a balloon or an airplane in the right place" to measure cloud electrical fields, although researchers are having more success with such attempts than in the past.

Scientists have proposed many hypotheses about what causes lightning, including the possibility that cosmic rays could supply high-energy electrons to initiate lightning. Uman wonders about these ideas. "It's a lot like astronomy. If there are things you can't really measure or measure well, you can do all kinds of speculation. Once you start accurate and thorough measurements, that puts a

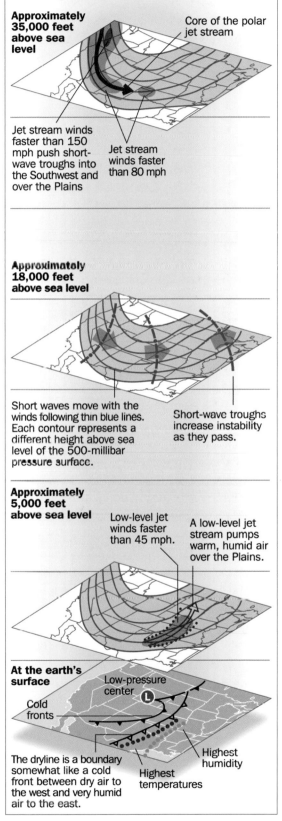

Tornado ingredients

The illustration here shows the factors that helped spawn sixty-six tornadoes in Oklahoma and Kansas on May 3 and 4, 1999. Similar patterns prompt forecasters to issue tornado watches because they make supercells and tornadoes likely by increasing atmospheric instability and adding spinning motions to the air.

Approximately 35,000 feet above sea level

Core of the polar jet stream

Jet stream winds faster than 150 mph push short-wave troughs into the Southwest and over the Plains

Jet stream winds faster than 80 mph

Approximately 18,000 feet above sea level

Short waves move with the winds following thin blue lines. Each contour represents a different height above sea level of the 500-millibar pressure surface.

Short-wave troughs increase instability as they pass.

Approximately 5,000 feet above sea level

Low-level jet winds faster than 45 mph.

A low-level jet stream pumps warm, humid air over the Plains.

At the earth's surface

Low-pressure center

Cold fronts

The dryline is a boundary somewhat like a cold front between dry air to the west and very humid air to the east.

Highest temperatures

Highest humidity

Meteorologists use the word "analysis" for a map or text describing weather patterns observed at a time in the past, usually the very recent past. A "prog," for prognosis, is a map or text describing the weather expected at a time in the future.

limit on it." The lightning lab at Camp Blanding is conducting some of the experiments that might help confirm or toss out the hypothesis that cosmic rays initiate lighting. Uman's doubts about the hypothesis are a piece of the scientific process. Experimental facts need to convince doubters before scientists accept a hypothesis.

Tornado ingredients

Since lightning and tornadoes both come from thunderstorms, researchers wonder whether lightning data from the Vaisala National Lightning Detection Network or from satellites could help forecast tornadoes. In some cases, lightning flashes have increased before a tornado formed, but this doesn't always happen. For now, tornado forecasters can rely on the Lightning Detection Network to help show them where thunderstorms are located, but not to indicate when a tornado could be forming. Like the NWS forecasters who predicted the tornado that destroyed the Parsons Company factory, meteorologists rely on other kinds of data to issue tornado warnings.

The first indication that severe thunderstorms or tornadoes are likely to threaten any part of the contiguous forty-eight U.S. states usually comes in an "outlook" produced by the NWS Storm Predic-

tion Center in Norman, Oklahoma. The Storm Prediction Center issues watches, which indicate that conditions are ripe for severe thunderstorms or tornadoes. Local NWS offices, such as the one at Lincoln, Illinois, issue warnings when someone spots and reports a severe thunderstorm or tornado, or when weather radar indicates that one is probably forming or has formed.

Walk into the Center in Norman, Oklahoma, and you'd see the computer screens and keyboards found in all weather offices today. Look a little closer at what's going on and you're likely to see a meteorologist doing something a Weather Bureau forecaster from a century ago would recognize: drawing lines on a paper map.

"We're one of the few places that still does hand analysis of weather maps," says Joe Schaefer, the Center's director. Why draw weather features by hand when computers churn out maps with the lines showing areas of high and low air pressure and fronts? "When you do this you see the data," Schaefer says, pointing to the lines forecaster Jack Hales is drawing on a map. "If you let the computer do it, you see [only] the line."

"You keep grounded in the data," Hales says. "The key to forecasting is deciding what's important and what's not," he says. Taking at least a few seconds to ponder questions, such as exactly where across Oklahoma the line designating a cold front should go, helps a forecaster see what data are worth more attention.

Storm Prediction Center alerts. In its three-day outlook on July 11, 2004, the Storm Prediction Center included Woodford County, Illinois (where the Parsons Company factory is located), in the area with a "slight" risk of severe thunderstorms on July 13. On the two-day outlook issued on July 12, the Center raised the risk level for July 13 to "moderate" in a large area including Woodford County.

On both days, the Lincoln NWS office passed along to the media and the general public the hazardous weather outlooks indicating the possibility of severe thunderstorms on July 13. At least a few of those who work at the Parsons Company factory probably watched a television meteorologist mention that the next day could bring a risk of severe thunderstorms. The most anyone could have done that evening was plan on staying alert to the next day's weather.

Those who were awake to watch the sun rise through the fog around 5:40 a.m. in central Illinois on July 13 saw nothing obvious to fear in the mild, humid weather. Some, especially those with weather radios, heard the Lincoln NWS office's 6 a.m. general forecast that mentioned the potential for strong thunderstorms with associated widespread wind damage and even tornadoes. Again, the only thing to do then was to vow to stay attentive to the sky and the weather radio and be ready to act quickly if a tornado is seen or a tornado warning is heard.

As the day began, an inversion, with temperatures above the ground warmer than at ground level, covered central Illinois. Such an inversion is very stable; it acts like a cap that keeps heated, ground-level air from rising. When the Lincoln NWS launched its regular 7 a.m. weather balloon, the ground temperature was 67 degrees, but the balloon's radiosonde found an inversion began only seventy feet above the ground where the temperature was 75 degrees. Temperatures aloft were warmer than at the ground up to 5,300 feet. Such

soundings are normal for the early morning when the sky is clear. The sun hasn't yet had time to heat the ground, which had cooled overnight.

Such a cap inhibits air from rising to form thunderstorms for a while. But it can help create stronger storms later in the day. By preventing air from rising, it allows low-level temperatures to grow hotter than they otherwise might while temperatures aloft change little. This means that when the cap disappears, even more energy is available to power thunderstorms. This is what happened across much of the Midwest on July 13, 2004.

At 10:52 a.m. Central Time, the Storm Prediction Center issued a tornado watch covering parts of Iowa and Illinois, including Woodford County.

A growing threat. When daylight time is in effect, NWS offices in the U.S. Central Time Zone launch weather balloons at 7 a.m. and 7 p.m. because these times are noon and midnight Coordinated Universal Time, respectively, when stations around the world launch balloons.

On July 13, 2004, the central Illinois weather

Cold air flowing out of a thunderstorm more than 100 miles away created this roll cloud north of Omaha, Nebraska. As the cold air pushed under very humid air, water vapor in the rising air condensed to form the cloud. Its bottom was approximately 400 feet above the ground and the top 800 feet up. The cold air was moving east at 35 mph and the temperature dropped from 67 to 39 degrees as it arrived. Such outflows sometimes trigger new thunderstorms.

Winds with names

Most of the time the wind is just "the wind," but some winds other than storm winds have such noticeable effects that people have given them names. We look at a few such winds on these two pages.

Land and sea breezes

On sunny days, land warms up quickly while the temperature of a nearby ocean or large lake changes little. At night, when the land cools, the ocean stays warm. The resulting temperature differences set up a pattern of different day and night local winds. Strong winds from larger weather systems overpower these local winds.

A daytime sea breeze

1 As land heats up, it warms the overlying air, which creates relatively low pressure at the surface but relatively higher pressure at any height aloft, such as 3,000 feet.

5 Air aloft flows from high pressure toward low pressure.

4 A line of clouds often forms along this sea breeze front.

2 The relatively cool ocean chills the air above it, which creates relatively high pressure at the surface but relatively lower pressure at any height aloft, such as 3,000 feet.

3 The pressure gradient force pushes air inland, creating a miniature, weak cold front.

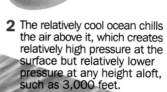

A nighttime land breeze

1 After sunset, the land begins cooling to create relatively high pressure at the surface.

4 Air aloft flows inland.

2 Air flows out to sea as a land breeze.

3 The ocean, which doesn't cool, warms the air above, creating relatively low pressure at the surface and relatively higher pressure at any height aloft, such as 3,000 feet.

If conditions are right, the miniature cold front can trigger nocturnal thunderstorms over the ocean.

When sea breezes converge

Converging Atlantic Ocean and Gulf of Mexico sea breezes, or sea breezes from both coasts converging with breezes from Lake Okeechobee, help trigger the numerous summer thunderstorms that make Florida the "lightning capital" of the United States. Similar converging breezes create clouds, showers, and thunderstorms in other parts of the world, such as in the southwestern part of Canada's Ontario Province.

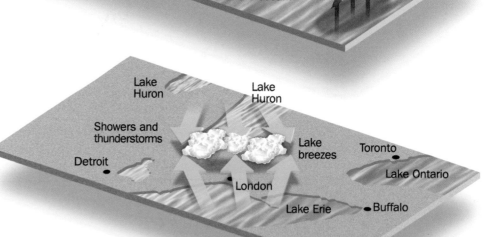

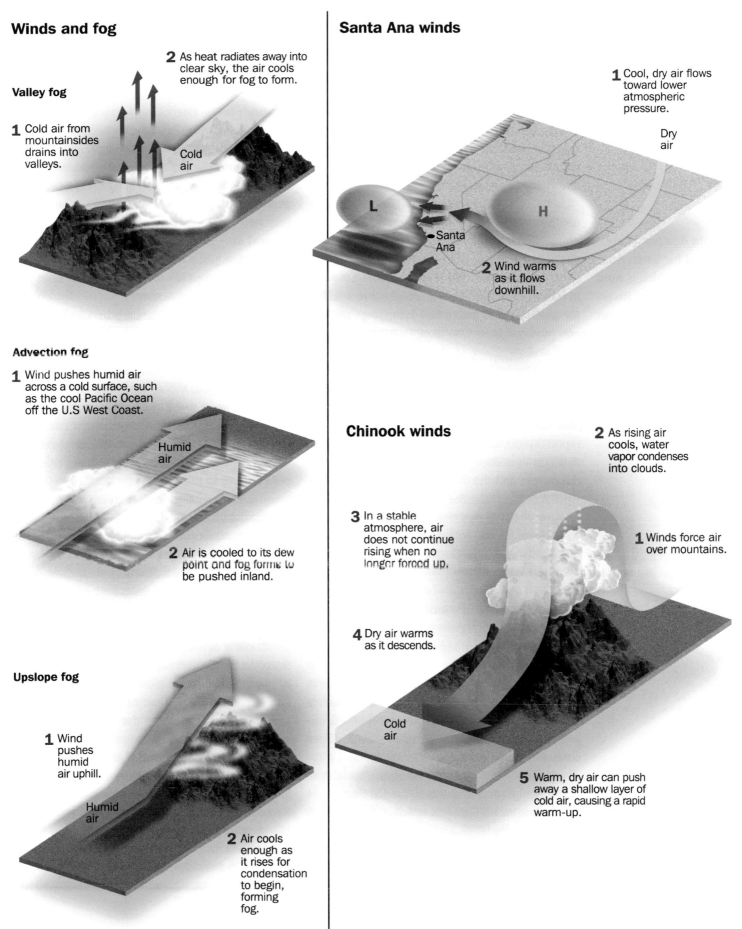

Winds and fog

Valley fog

1 Cold air from mountainsides drains into valleys.

2 As heat radiates away into clear sky, the air cools enough for fog to form.

Cold air

Advection fog

1 Wind pushes humid air across a cold surface, such as the cool Pacific Ocean off the U.S West Coast.

Humid air

2 Air is cooled to its dew point and fog forms to be pushed inland.

Upslope fog

1 Wind pushes humid air uphill.

Humid air

2 Air cools enough as it rises for condensation to begin, forming fog.

Santa Ana winds

1 Cool, dry air flows toward lower atmospheric pressure.

Dry air

L

H

●Santa Ana

2 Wind warms as it flows downhill.

Chinook winds

2 As rising air cools, water vapor condenses into clouds.

3 In a stable atmosphere, air does not continue rising when no longer forced up.

1 Winds force air over mountains.

4 Dry air warms as it descends.

Cold air

5 Warm, dry air can push away a shallow layer of cold air, causing a rapid warm-up.

213

Finding patterns in satellite images

In the 1970s when Robert Maddox was working at the NOAA Laboratory in Boulder, Colorado, and earning a PhD at Colorado State University in Fort Collins, weather satellites began offering hope of solving problems he had faced as an Air Force forecaster in the 1960s and 1970s.

During his eight years in the Air Force, Maddox used weather radars that showed "just white blobs" where precipitation was falling but no de-

Robert Maddox in front of an enlarged infrared satellite image of a mesoscale convective complex.

tails. Forecasters were all but blind at night because satellite infrared images "showed you where the clouds were but little else."

When satellite images began showing cloud-top temperatures in the late 1970s, Maddox realized they could help solve many of the problems he had faced as a forecaster. "I could eyeball [the images] day after day and keep track of what was going on in the upper air." Until Maddox began tracking the infrared images, most meteorologists had assumed that summertime thunderstorms were random. "The thing that struck me was the coherent structure of the [cloud top] temperatures. It wasn't a cold blob here, a cold blob there. What really astounded me is how frequently [organized systems of thunderstorms] occurred."

Maddox credits his Colorado State PhD committee, "especially Professor Bill Cotton and my NOAA colleague J. Michael Fritsch" with encour-

aging him to study what he later named mesoscale convective complexes for his PhD dissertation. He says Cotton and Fritsch and their students provided invaluable help with research on MCC global climatologies, their internal structures, and rainfall production.

Maddox traces his interest in weather to fishing and golfing with his father when he was growing up in Granite City, Illinois. "I spent most teenage years outside, which led me to become a sky and cloud watcher. This led to an interest in thunderstorms and attempts to make my own storm forecasts when we'd head for the golf course."

After graduating from high school in 1962, when high school teachers were encouraging every one with good grades to become an engineer, I headed off to Purdue to major in engineering," he says. "But during my freshman year found it boring. It dawned on me that what I was really interested in was the weather." He transferred to Texas A&M University to study meteorology.

Maddox says he also grew up as "an avid reader in many genres." The accurate weather and storm descriptions and use of Navajo weather legends and myths in Tony Hillerman's mysteries built around two Navajo tribal policemen "got me to wondering about how the Plains Indians' folklore dealt with tornadoes and large hailstorms." While he found "almost no severe storm mythology in the records about Plains tribes," his extensive reading made him a student of Plains history and folklore.

Now retired, Maddox is turning to an extensive study with a half dozen other meteorologists of the Tri-State Tornado of March 18, 1925. It followed a 219-mile path from southeastern Missouri, across southern Illinois, and into southwestern Indiana, killing 695 people. It's the deadliest tornado in U.S. history and has the longest path on record.

"I am leading a reanalysis of the meteorological conditions," Maddox says.

"We have had some success, and I'm starting to do a new series of map plots incorporating all the data we've dredged up. Others are preparing a detailed documentation of the entire damage path. "I think that eventually we will document the meteorology and the tornado path far better than anything that has ever appeared in the literature."

Mesoscale convective complex

This illustration is a simplified view of the main features of a typical mesoscale convective complex.

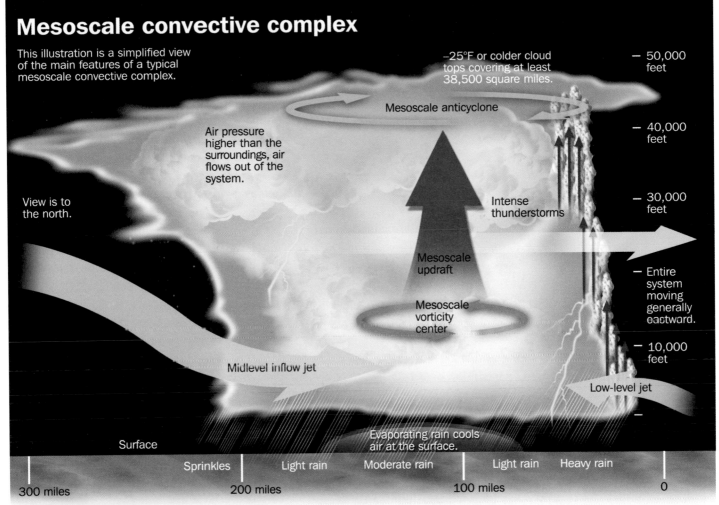

When the low-level jets dies in the early morning the rain ends and clouds begin evaporating. The mesoscale vorticity center continues rotating as it moves east where it can help a new MCC form the next evening.

In the 1980 *BAMS* article, Maddox identified forty-three MCCs from the night of March 22–23, 1978, through the night of August 25–26 that year.

The criteria for a mesoscale convective complex that he used in the article—which are sometimes still used—were:

- Circular or only slightly elliptical area of cloud tops. (This criterion ensures that squall lines are not counted as MCCs.)
- Cloud-top temperatures −25 degrees Fahrenheit or colder covering at least 100,000 square kilometres (38,627 square miles or about the size of Iowa).
- A center area of cloud-top temperatures −62 degrees Fahrenheit or colder covering at least 50,000 square kilometres (19,313 square miles). This indicates that deep convection is ongoing.

- The system lasts at least six hours.

Maddox found that "severe thunderstorm phenomena (tornadoes, wind, and hail), as well as torrential rains and flash floods, were often associated with the systems," with one in five causing injuries or deaths and seventeen of the forty-three producing heavy rain.

Maddox found that three key ingredients are needed for an MCC to form:

- A weak short wave (short waves are described in Chapter 5) approximately 18,000 feet above the ground, to help organize the system.
- A strong low-level jet, which we will discuss later in this chapter.
- Substantial CAPE south and southwest of where the system forms. (CAPE is explained in the "Powering Thunderstorms" section of Chapter 8.)

Lake-effect snow

Frigid air blowing across relatively warm water creates the lake-effect snow that makes places around the Great Lakes unusually snowy. Such snow sometimes falls near the Great Salt Lake in Utah and on a few places near oceans, including in Japan and Scandinavia.

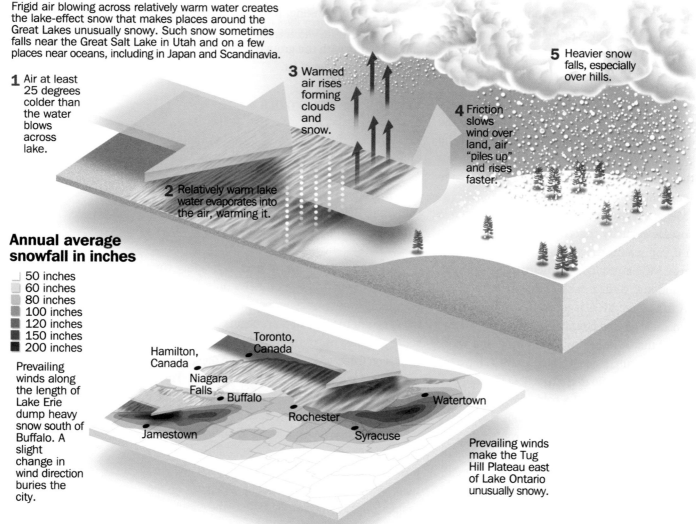

1 Air at least 25 degrees colder than the water blows across lake.

2 Relatively warm lake water evaporates into the air, warming it.

3 Warmed air rises forming clouds and snow.

4 Friction slows wind over land, air "piles up" and rises faster.

5 Heavier snow falls, especially over hills.

Annual average snowfall in inches

- 50 inches
- 60 inches
- 80 inches
- 100 inches
- 120 inches
- 150 inches
- 200 inches

Prevailing winds along the length of Lake Erie dump heavy snow south of Buffalo. A slight change in wind direction buries the city.

Hamilton, Canada
Toronto, Canada
Niagara Falls
Buffalo
Jamestown
Rochester
Syracuse
Watertown

Prevailing winds make the Tug Hill Plateau east of Lake Ontario unusually snowy.

Virginia. Meteorologists call such a layer of warm air between layers of cold air an elevated warm layer, or a melting layer. They know it's a sign of trouble.

By noon on December 15, the weather at each of the cities on the map was:

- Panama City, Florida: 68 degrees, cloudy, no precipitation
- Athens, Georgia: 34 degrees, rain
- Hickory, North Carolina: 32 degrees, freezing rain
- Roanoke, Virginia: 27 degrees, ice pellets (sleet)

At Athens, snow could have been falling from high clouds, but it melted into raindrops in the relatively thick layer of warm air.

Farther north, over Hickory, snow melted as it fell into the layer of warm air aloft but began cooling as it fell through a thicker layer of cold air to the south. The drops cooled below 32 degrees, becoming supercooled but remaining liquid, and instantly turned to ice when they hit trees, power lines, and roads.

Even farther north, over Roanoke, the layer of warm air was thinner and the layer of cold air right above the ground was thicker than farther south. Snow falling into the warm air began melting but didn't completely melt. As the partly melted ice crystals fell through the cold air, they froze into tiny pieces of ice, about the size of the tip of a pencil lead. The NWS reports such precipitation as "ice pellets," but most people in the United States call it sleet. (A mixture of freezing rain and sleet is common and some people refer to this as "sleet.")

The real weather, of course, is more complicated than the idealized picture in the graphic. For example, the Roanoke weather station reported light freezing rain from 5:35 a.m. until 6:23 a.m. Snow began mixing with the freezing rain and then it turned entirely to snow two minutes later. Snow continued until 8:56 a.m. when freezing rain began mixing with it. Ice pellets (sleet) fell from 10:54 a.m. through 2:01 p.m. Freezing rain then fell again until 7:54 p.m., when the precipitation ended. "Wintry mix" is a good description of Roanoke's weather that day.

During the day, as much as three-quarters of an inch of ice accumulated on parts of northern Georgia and South Carolina, western North Carolina, and southwestern Virginia. The ice pulled down tree limbs and power lines, causing almost 700,000 customers to lose power, as well as causing numerous accidents on icy roads.

When freezing rain lasts several hours and deposits more than a quarter inch of ice, the NWS calls the event an **ice storm**. Forecasts of such events prompt ice storm warnings.

When ice crystals form in cold clouds and fall to the ground through below-freezing air, they reach the ground as snow, and when large amounts fall it can cause major disruptions. However, only an inch of freezing rain that coats everything with ice can cause more delays and damage than a foot of fluffy snow.

Atmospheric rivers

In the opening of Chapter 4, we joined scientists on a NOAA WP-3 research airplane off the California coast in February 1998 who were learning more about what happens as an "atmospheric river" carrying huge amounts of water vapor from the tropics reaches the northern California coast.

The follow-up research question is: What happens to all of that water vapor? In simple terms, when a concentrated blast of very humid air hits North America's west coast, the wind pushes the humid air up over the mountains; the air cools, water vapor condenses, and rain or snow falls. Exactly how much rain falls and where it falls, or how much snow falls and where it falls, are mesoscale meteorology questions that researchers and forecasters are trying to answer.

Answering these questions is important, not only because of the flood danger, but also because

A match for the Mississippi

Martin (Marty) Ralph, one of the NOAA scientists who led the 1998 research flight, says that the atmospheric river that he and his colleagues described in a 2004 journal article was transporting more than 50 million kilograms (17 million pounds) of water vapor to the California coast each second. In terms of liquid water, "a typical atmospheric river transports about three times what the Mississippi River transports and about one-third of what the Amazon River transports. Fortunately for those who live in flood-prone areas, Ralph says, typically only 1 to 10 percent of the water vapor in the air falls as precipitation.

water needs to be managed in the western United States. Most of the West has a wet season during the cool part of the year and a dry season during the warm months. The contrast is greatest in the mountains, as shown in the chart below of average monthly precipitation (melted snow and rain) and snow for Blue Canyon, California. Blue Canyon is a mile above sea level, just off Interstate 80 east of Sacramento as the highway climbs the Sierra Nevada on the way to Reno, Nevada.

Blue Canyon average monthly precipitation and snowfall (in inches)

Month	Precipitation	Snowfall
January	13.19	50.9
February	10.47	44.6
March	9.40	52.5
April	5.32	26.6
May	3.21	7.8
June	0.90	0.7
July	0.24	0.0
August	0.42	0.0
September	1.02	0.4
October	3.62	2.9
November	9.37	24.6
December	12.71	41.1

Snow that falls in the Sierra Nevada, the generally north-south mountain range in eastern California, supplies about 35 percent of the water used all year by all of the state. Good water management requires accurate data on how much precipitation is falling, how much of it is snow, and how much is rain.

Inside a hurricane

Satellite images make hurricanes look relatively simple: A swirl of clouds wrapped around a clear eye. In reality, a hurricane is an incredibly complex mixture of swirls within swirls, rising and sinking air, energy exchanges, and reactions back and forth between the storm and the surrounding air and the underlying ocean and land.

Major hurricane features and influences

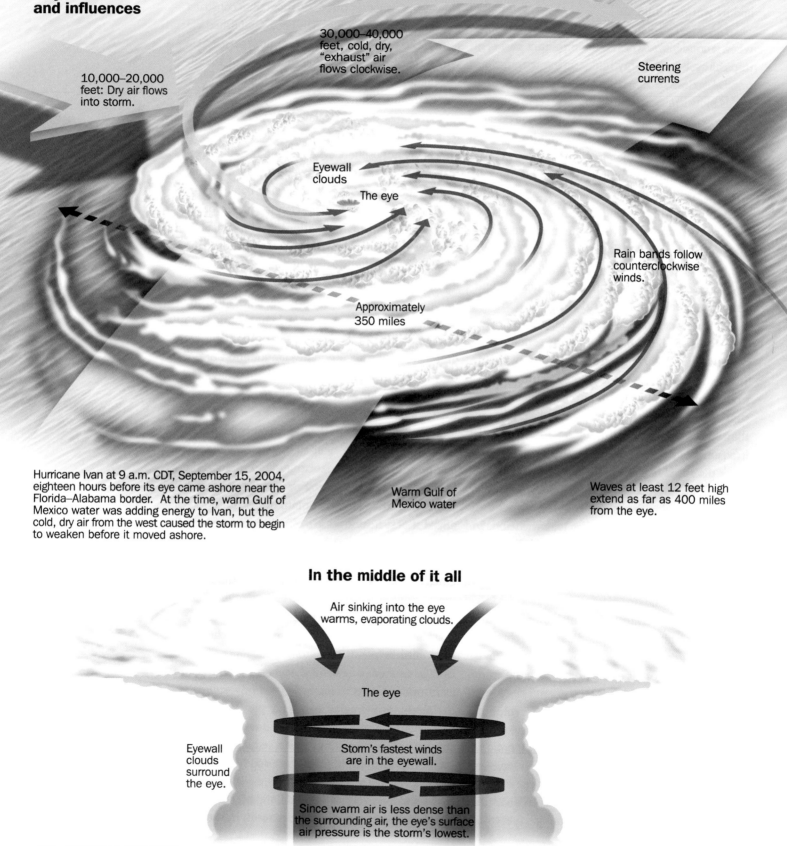

Dry air sinks and warms, evaporating clouds to create the clear weather that often precedes or follows a hurricane.

30,000–40,000 feet, cold, dry, "exhaust" air flows clockwise.

10,000–20,000 feet: Dry air flows into storm.

Steering currents

Eyewall clouds

The eye

Rain bands follow counterclockwise winds.

Approximately 350 miles

Hurricane Ivan at 9 a.m. CDT, September 15, 2004, eighteen hours before its eye came ashore near the Florida–Alabama border. At the time, warm Gulf of Mexico water was adding energy to Ivan, but the cold, dry air from the west caused the storm to begin to weaken before it moved ashore.

Warm Gulf of Mexico water

Waves at least 12 feet high extend as far as 400 miles from the eye.

In the middle of it all

Air sinking into the eye warms, evaporating clouds.

The eye

Eyewall clouds surround the eye.

Storm's fastest winds are in the eyewall.

Since warm air is less dense than the surrounding air, the eye's surface air pressure is the storm's lowest.

The hurricane heat engine

Like the engine in your car, a hurricane is a heat engine that converts potential energy into motion: wind, which in turn powers waves and storm surge. The images below are an idealized look at a hurricane's energy flow.

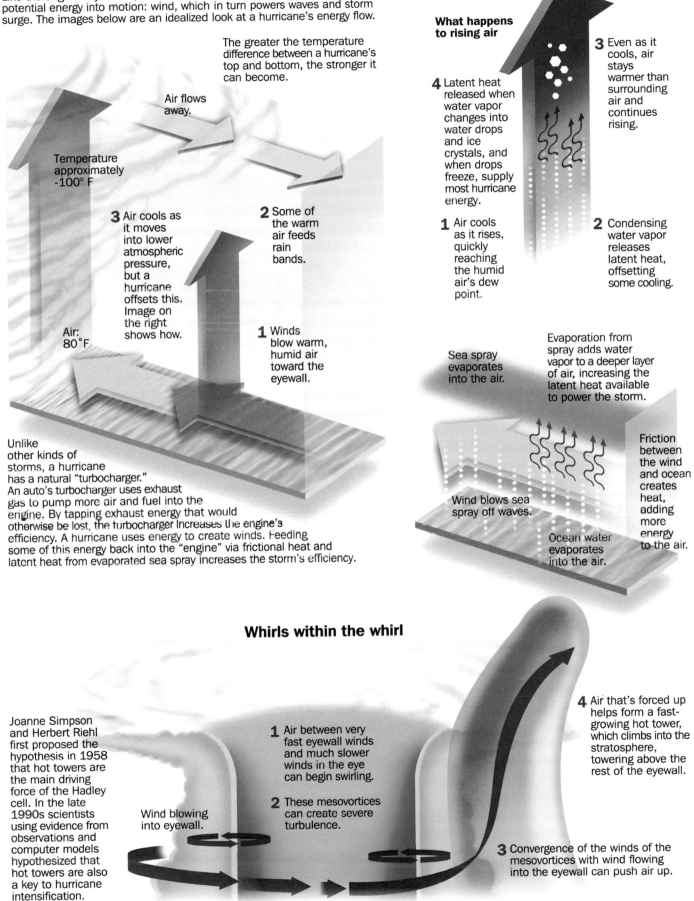

The greater the temperature difference between a hurricane's top and bottom, the stronger it can become.

Air flows away.

Temperature approximately -100° F

3 Air cools as it moves into lower atmospheric pressure, but a hurricane offsets this. Image on the right shows how.

2 Some of the warm air feeds rain bands.

Air: 80°F

1 Winds blow warm, humid air toward the eyewall.

Unlike other kinds of storms, a hurricane has a natural "turbocharger." An auto's turbocharger uses exhaust gas to pump more air and fuel into the engine. By tapping exhaust energy that would otherwise be lost, the turbocharger increases the engine's efficiency. A hurricane uses energy to create winds. Feeding some of this energy back into the "engine" via frictional heat and latent heat from evaporated sea spray increases the storm's efficiency.

What happens to rising air

3 Even as it cools, air stays warmer than surrounding air and continues rising.

4 Latent heat released when water vapor changes into water drops and ice crystals, and when drops freeze, supply most hurricane energy.

1 Air cools as it rises, quickly reaching the humid air's dew point.

2 Condensing water vapor releases latent heat, offsetting some cooling.

Evaporation from spray adds water vapor to a deeper layer of air, increasing the latent heat available to power the storm.

Sea spray evaporates into the air.

Wind blows sea spray off waves.

Ocean water evaporates into the air.

Friction between the wind and ocean creates heat, adding more energy to the air.

Whirls within the whirl

Joanne Simpson and Herbert Riehl first proposed the hypothesis in 1958 that hot towers are the main driving force of the Hadley cell. In the late 1990s scientists using evidence from observations and computer models hypothesized that hot towers are also a key to hurricane intensification.

Wind blowing into eyewall.

1 Air between very fast eyewall winds and much slower winds in the eye can begin swirling.

2 These mesovortices can create severe turbulence.

4 Air that's forced up helps form a fast-growing hot tower, which climbs into the stratosphere, towering above the rest of the eyewall.

3 Convergence of the winds of the mesovortices with wind flowing into the eyewall can push air up.

239

Ingredients needed for a hurricane

Hurricanes and other tropical cyclones form over tropical oceans and the western subtropical Atlantic Ocean because these regions have the needed ingredients.

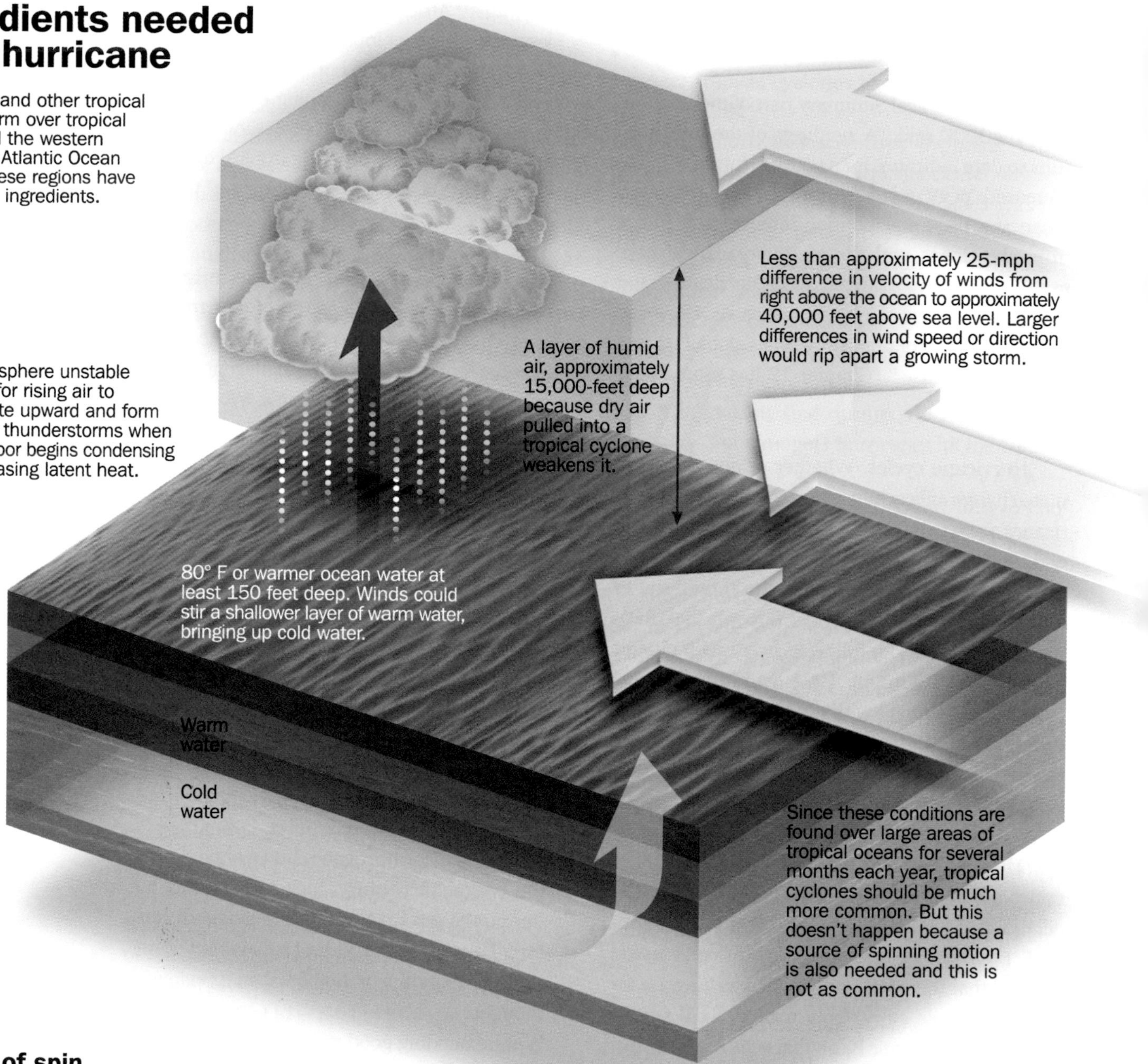

An atmosphere unstable enough for rising air to accelerate upward and form towering thunderstorms when water vapor begins condensing and releasing latent heat.

A layer of humid air, approximately 15,000-feet deep because dry air pulled into a tropical cyclone weakens it.

Less than approximately 25-mph difference in velocity of winds from right above the ocean to approximately 40,000 feet above sea level. Larger differences in wind speed or direction would rip apart a growing storm.

80° F or warmer ocean water at least 150 feet deep. Winds could stir a shallower layer of warm water, bringing up cold water.

Warm water

Cold water

Since these conditions are found over large areas of tropical oceans for several months each year, tropical cyclones should be much more common. But this doesn't happen because a source of spinning motion is also needed and this is not as common.

Sources of spin

Decaying systems that move over tropical oceans from the middle latitudes, such as fronts and mesoscale convective complexes, supply the spin for many tropical cyclones. But 60% of Atlantic tropical storms and 85% of the major hurricanes begin with easterly waves that move across Africa and across the Atlantic Ocean.

An easterly wave

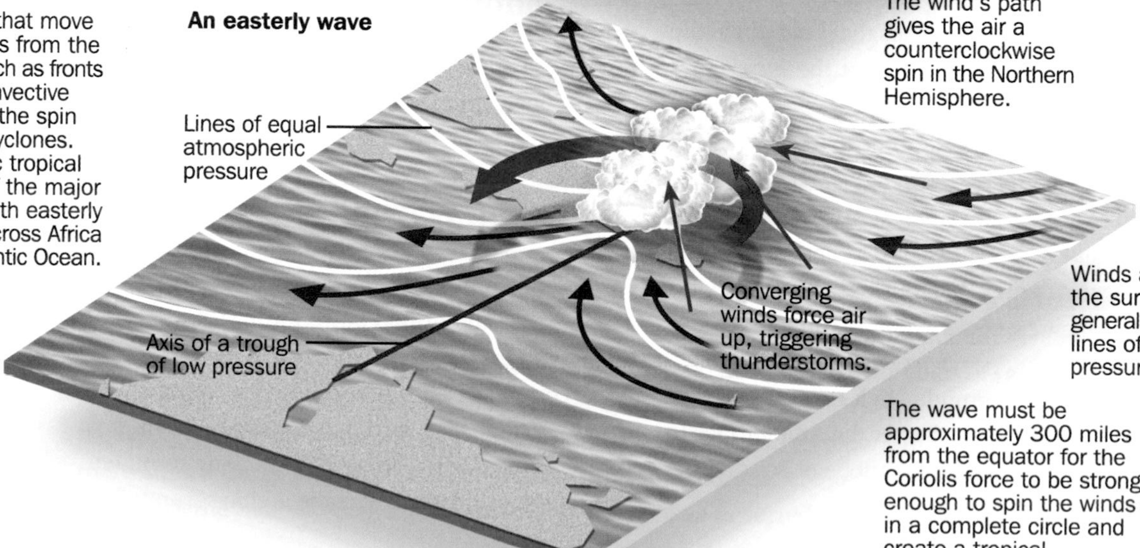

Lines of equal atmospheric pressure

Axis of a trough of low pressure

Converging winds force air up, triggering thunderstorms.

The wind's path gives the air a counterclockwise spin in the Northern Hemisphere.

Winds above the surface generally follow lines of equal pressure.

The wave must be approximately 300 miles from the equator for the Coriolis force to be strong enough to spin the winds in a complete circle and create a tropical depression.

miles of Miami twenty-six times a century. Globally, fewer than 100 tropical cyclones form each year.

One of the required conditions is that winds take on the spinning motion caused by the Coriolis force. This is why tropical cyclones usually don't form within about five degrees of latitude of the equator—the Coriolis force is zero at the equator and increases as you head toward the poles. Too far toward the poles, however, and cooler water or unfavorable winds keep tropical cyclones from forming.

The eye

The calm eye at the center of a mature tropical cyclone distinguishes it from almost all other types of storms, and eye formation is usually a good sign that a tropical storm has reached the threshold of becoming a hurricane with highest sustained winds of at least 74 mph. When a tropical cyclone is strongest, an eyewall of towering clouds containing the storm's fastest winds surrounds the eye, which is generally five to twenty miles across. Winds are calm, or nearly calm, in the eye, and it is often cloud free, or nearly cloud free.

Edward R. Murrow, the famous radio and television reporter of the 1940s and 1950s, shared his experience of flying through the eyewall into the eye of a hurricane in a 1954 CBS *See It Now* documentary. The flight in an Air Force WB-29 went into Hurricane Edna a couple of days before it hit New England.

In the documentary, we see Murrow sitting in the four-engine bomber's glass nose with a view in all directions. "It's like flying through milk," Murrow says as the airplane flies through the eyewall with little turbulence.

You hear a sharp bang and Murrow says, "We've hit something with a bang, audible above the roar of the motors; a solid sheet of water."

Then Murrow says, "Nose is down, now we're going up again. A mite rough out here now… There's the sun, there it is. I can't see the eye yet." We hear shouts from the crew, and Murrow says, "What a beautiful sight. We're in an amphitheater surrounded by clouds. [The ocean] looks like a lovely alpine lake surrounded by snow."

Later, back in the studio, Murrow reflected on his experience. "The eye of a hurricane is an excellent place to reflect upon the puniness of man and

The unique Atlantic Basin
The Gulf Steam helps make the Atlantic Ocean warm enough for tropical cyclones to form and continue at higher latitudes than in the other basins. This makes Atlantic storms more likely to be weakened or killed by wind shear, which can interfere with the flow of energy in the storm. As you head north or south from the equator, wind shear is more prevalent than in the deep tropics. The larger role of wind shear is one reason that the number of Atlantic Basin tropical cyclones varies more from year to year than in other basins.

his works. If an adequate definition of humility is ever written, it's likely to be done in the eye of a hurricane."

Even though airplanes have been flying into the eyes of tropical cyclones since 1944, and scientists have been conducting focused research from airplanes inside hurricanes since 1957, "the exact mechanism by which the eye forms remains somewhat controversial," according to the Frequently Asked Questions section of the NOAA Hurricane Research Division Web site.

What is known is that some of the air that rises to the cyclone's top in the eyewall sinks into the center of the eye and warms as it descends, with the greatest warming relative to the surrounding air occurring in the top of the eye. Since the eye is warmer than the surrounding eyewall, and air becomes less dense as it warms, a tropical cyclone's lowest pressure is in the eye.

While a hurricane's lowest pressure is in the eye, its strongest winds and tallest thunderstorms are in the eyewall, which surrounds the eye. The **rain bands** that spiral in toward the eye are generally made of smaller thunderstorms with weaker winds.

In a strong hurricane, the eyewall surrounds the eye, but as a storm begins to weaken, part of the eyewall can disappear. Also, sometimes a second, larger eye forms around the original eye, which then dissipates. As we saw above, on August 27, 2005, Jack Beven at the National Hurricane Center was watching Katrina go through such an eyewall replacement cycle. We examine this below in the section on "Intensity Forecasting."

Hurricane flights

Much of what we know today about tropical cyclones we learned from airplane flights into tropi-

Intense lows that form over polar oceans sometimes develop eyes and have 74 mph or faster hurricane-force winds. Satellite images show that such storms sometimes have warm cores like tropical cyclones.

Al Roker, weatherman for NBC's *Today Show*, views the eye of Hurricane Dean from the flight deck of an Air Force Reserve WC-130 on August 19, 2007. Calm winds in the eye make it safe to stand even though winds in the surrounding eyewall were estimated at 155 mph, just short of making Dean a Category 5 hurricane.

cal cyclones beginning during World War II. The flights continue to enable meteorologists to make full use of technologies such as radar, satellites, and computer models. Flights give scientists **ground truth** data that help them understand radar and satellite images and test models.

For scientists, flying into a hurricane provides more than data, says Hugh Willoughby, who has flown into more than 400 tropical cyclones beginning with typhoon flights in the 1970s as a Navy weather officer. He headed NOAA's Hurricane Research Division from 1995 to 2002 before becoming a distinguished research professor at Florida International University. "I find that what we see in the air is like looking in the back of the book for the answer. Casual visual observations give you hints about what's going on," he says. "Flying into a storm restores a balance [when] I'm lost in the equations."

Observational flights into tropical cyclones began during World War II to find and track typhoons and hurricanes that endangered airplanes, navy ships, and vital factories near the coasts. People began calling military aviators who flew into storms during World War II "hurricane hunters," an appropriate name before the advent of weather satellites, because flights often went out to locate a storm meteorologists suspected was somewhere over the horizon. Until the United States started launching weather satellites in the 1960s, forecasters relied on patterns of clouds and ocean waves for signs that a hurricane had formed. The "hurricane hunter" name has stuck even though today's

hurricane fliers no longer have to hunt for hurricanes.

In 1944, the Weather Bureau credited data from flights into what became known as the "Great Atlantic Hurricane" with the vastly improved forecasts that helped keep the death toll on land low —fifty instead of the 600 deaths a similar hurricane caused on Long Island and across New England in 1938. (At sea the 1944 hurricane sank a Navy destroyer and three Coast Guard ships, killing 298 men. The call of wartime duty and some bad decisions put the ships in the storm's path.) The success of the 1944 flights guaranteed the future of the hurricane hunters.

Edward R. Murrow's comment about Hurricane Edna in his 1954 hurricane flight documentary sums up the fate of virtually every hurricane that's threatened the United States since late in World War II: Its "movements were reported as completely as those of a president or a movie star."

Flight data needed. Despite the advances in weather radar and satellites, remote sensing can't supply all of the data National Hurricane Center forecasters need, especially as a strong hurricane approaches land. As we saw in Chapter 6, weather radar waves do not follow the earth's curve, which means that 140 miles from the radar antenna, the radar detects only what's happening a mile and a half above the ground or higher. The hurricane winds that forecasters need to measure are those immediately above the ocean, which will blow down trees and buildings when the hurricane moves ashore.

Though weather satellites can't directly measure a tropical cyclone's winds and air pressures, all of the world's tropical cyclone forecasting centers, except for those at the U.S. National Hurricane Center (NHC), rely heavily on satellite images to estimate the strength of tropical cyclones. In fact, the NHC uses satellite estimates of storm strength when storms are far out over the Atlantic Ocean or over the eastern Pacific. When a storm threatens Hawaii, California, or Mexico's Pacific coast, the NHC can call for airplane reconnaissance, as it did in September 2006 for Hurricane Lane, which spent all four days of its short life close to Mexico's West coast. Lane threatened cities such as Acapulco and Los Cabos before coming ashore as a small, Category 3 hurricane with 125 mph winds in a

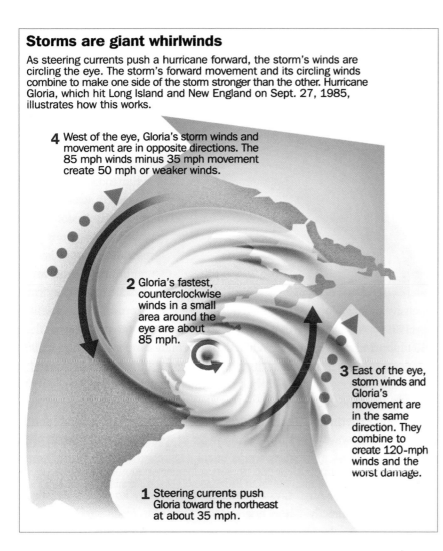

Storms are giant whirlwinds

As steering currents push a hurricane forward, the storm's winds are circling the eye. The storm's forward movement and its circling winds combine to make one side of the storm stronger than the other. Hurricane Gloria, which hit Long Island and New England on Sept. 27, 1985, illustrates how this works.

4 West of the eye, Gloria's storm winds and movement are in opposite directions. The 85 mph winds minus 35 mph movement create 50 mph or weaker winds.

2 Gloria's fastest, counterclockwise winds in a small area around the eye are about 85 mph.

3 East of the eye, storm winds and Gloria's movement are in the same direction. They combine to create 120-mph winds and the worst damage.

1 Steering currents push Gloria toward the northeast at about 35 mph.

Airplane wind measurements. A recent study that compared wind-speed estimates based primarily on satellites with airplane measurements from the same storms found that less than ten percent of the estimates were in error by more than 20 mph. Still, such an error could mean that the Hurricane Center could be off by an entire category for a storm that's approaching the coast. That's a risk that U.S. policy makers don't want to take.

Beginning with Katrina's formation near the Bahamas until it came ashore in Mississippi and Louisiana, at least one Air Force Reserve or NOAA airplane was in the storm almost all of the time.

sparsely populated part of the state of Sinaloa. Lane caused widespread damage and was blamed for four deaths.

Satellite wind estimates. In the late 1960s, as weather satellite technology was quickly improving, Vernon Dvorak, who was a scientist with the Environmental Sciences Services Administration (it became NOAA in 1970), was one of the scientists working on ways to use satellite images to estimate tropical cyclone wind speeds. Dvorak first described what is now called the Dvorak Technique in a 1972 NOAA technical memorandum. The technique continues to be regularly updated to take advantage of new satellite technologies and advances in storm knowledge. It uses not only the shapes of clouds, but also their temperature patterns, such as the differences in temperature in the eye and the tops of the clouds around the eye as detected by infrared sensors, to estimate storm strength.

The WC-130 airplanes from the Air Force Reserve 53rd Weather Reconnaissance Squadron based in Biloxi, Mississippi, conduct storm reconnaissance. Collected data include the exact location of the eye, surface air pressure, and distribution of winds, which Hurricane Center forecasters use in their predictions.

The NOAA WP-3 airplanes from NOAA's Aircraft Operations Center at MacDill Air Force Base in Tampa, Florida, have more sophisticated data collecting equipment than the Air Force airplanes. Their main task is gathering research data, but they also send regular reports to Hurricane Center forecasters. NOAA uses these airplanes for other projects, such as air quality research or the extratropical cyclone study that opened Chapter 5. Both NOAA and the Air Force Reserve airplanes sometimes collect data on winter storms, especially over the Pacific Ocean.

During Katrina and a few other 2005 hurricanes, a Naval Research Laboratory P-3 joined the

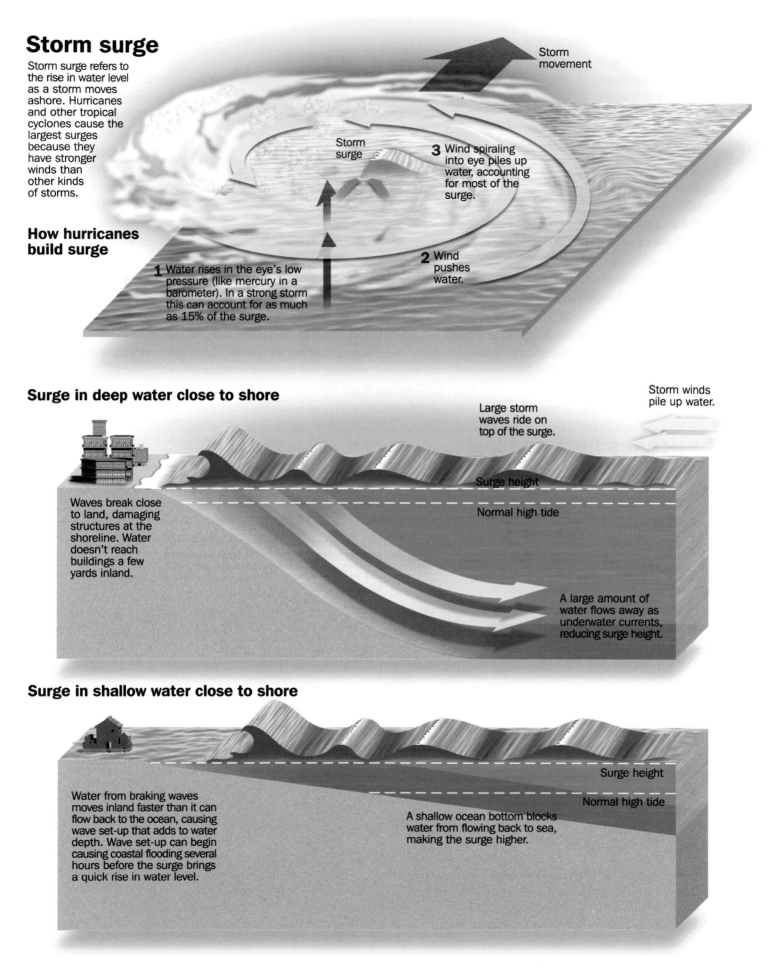

Storm surge

Storm surge refers to the rise in water level as a storm moves ashore. Hurricanes and other tropical cyclones cause the largest surges because they have stronger winds than other kinds of storms.

How hurricanes build surge

Storm movement

Storm surge

3 Wind spiraling into eye piles up water, accounting for most of the surge.

2 Wind pushes water.

1 Water rises in the eye's low pressure (like mercury in a barometer). In a strong storm this can account for as much as 15% of the surge.

Surge in deep water close to shore

Storm winds pile up water.

Large storm waves ride on top of the surge.

Surge height

Normal high tide

Waves break close to land, damaging structures at the shoreline. Water doesn't reach buildings a few yards inland.

A large amount of water flows away as underwater currents, reducing surge height.

Surge in shallow water close to shore

Surge height

Normal high tide

Water from braking waves moves inland faster than it can flow back to the ocean, causing wave set-up that adds to water depth. Wave set-up can begin causing coastal flooding several hours before the surge brings a quick rise in water level.

A shallow ocean bottom blocks water from flowing back to sea, making the surge higher.

malized Hurricane Damages in the United States: 1925–1995," published in the September 1998 issue of the AMS journal *Weather and Forecasting*.

They examined 244 hurricanes and tropical storms that hit the United States and found that the two Category 5 and ten Category 4 storms caused 49 percent of the hurricane-related property losses and that the forty Category 3 storms caused another 35 percent of the losses. The remaining 192 Category 1 and 2 hurricanes and the tropical storms caused only 16 percent of the property losses.

As we saw in the opening of this chapter, less than twelve hours before Katrina was expected to come ashore, James Franklin was trying to make sense of conflicting signals regarding its strength, which made forecasting its intensity the next morning difficult. During a storm, a forecaster has "a whole bunch of indirect measures (of wind speeds) and they often conflict," Franklin says. "It's kind of like putting together a jigsaw puzzle when the pieces all have fuzzy edges and they don't fit together quite right. You have to look at all these pieces and make it all fit, understanding the limitations and error, and try and put the puzzle together."

He notes that forecasting in some places, such as Japan, where he would have only satellite estimates of the winds, would be a lot easier, but the bulletins on current storm conditions and the forecasts wouldn't be as precise.

The difficulties of determining wind speeds begin with the fact that when forecasters talk about a Category 5 hurricane, they are not talking about a storm with winds blowing 155 mph or faster over a large area, such as completely around the eye, much less over the entire storm.

Measuring hurricanes. National Hurricane Center forecasters have access to a great deal of wind data for any storm that threatens land. In addition to satellite estimates, these include:

- Calculations of wind speed and direction by a WP-3's or WC-130's flight management system using the airplane's heading and air speed compared with its actual track and speed over the ocean.
- Measurements by **dropsondes**, which are packages of instruments that airplanes drop that radio data back as they descend to the ocean under small parachutes.

- Readings from the **stepped-frequency microwave radiometer (SFMR)** aboard the NOAA WP-3s and Air Force Reserve WC-130s.
- Data from ocean buoys.

Using any of these is far from straightforward. The airplane is flying some 10,000 feet above the ocean where winds are usually faster than at the surface. Forecasters usually use a reduction factor, such as assuming surface winds will be 90 percent as fast as those aloft, but this can differ not only from storm to storm but from hour to hour in the same storm.

The dropsondes give accurate measurements, but they are sampling winds in only the very small area they are falling through. A forecaster is never really sure whether a dropsonde's measurement right above the ocean was in a wind gust or in a lull between gusts. Still, they are actual wind speeds and they, along with buoy measurements, are used to calibrate both SFMR measurements and the reduction factor from flight level winds.

The stepped-frequency microwave radiometer gives an overall picture of the storm's surface winds, but it is still being refined. During Katrina, for instance, scientists including Franklin were testing improvements in the instrument.

As it turned out, the scientists involved said at a March 2006 hurricane conference that the "unprecedented hurricane frequency and intensity [in 2005] provided ample data to evaluate both instrument performance and wind retrieval quality in conditions up to…Saffir-Simpson Category 5." One result was a suggested change in the model used to relate instrument readings to surface wind speeds.

Wind speeds are related to the air pressure in a hurricane's eye; in general, the lower the pressure, the stronger the wind, and forecasters have some rules of thumb relating pressure to wind speed. But there are exceptions. As we learned in Chapter 3, the pressure gradient force depends not only on the pressure difference between high and low pressure but also on the distance between the areas of high and low pressure. If a storm spreads out, this distance could grow and the wind speed could drop even though the central pressure stays the same. In fact, says Hugh Willoughby, "Generally about one-third of maximum winds estimated from pressure-wind relationships are high or low by as much as half a Saffir-Simpson Category."

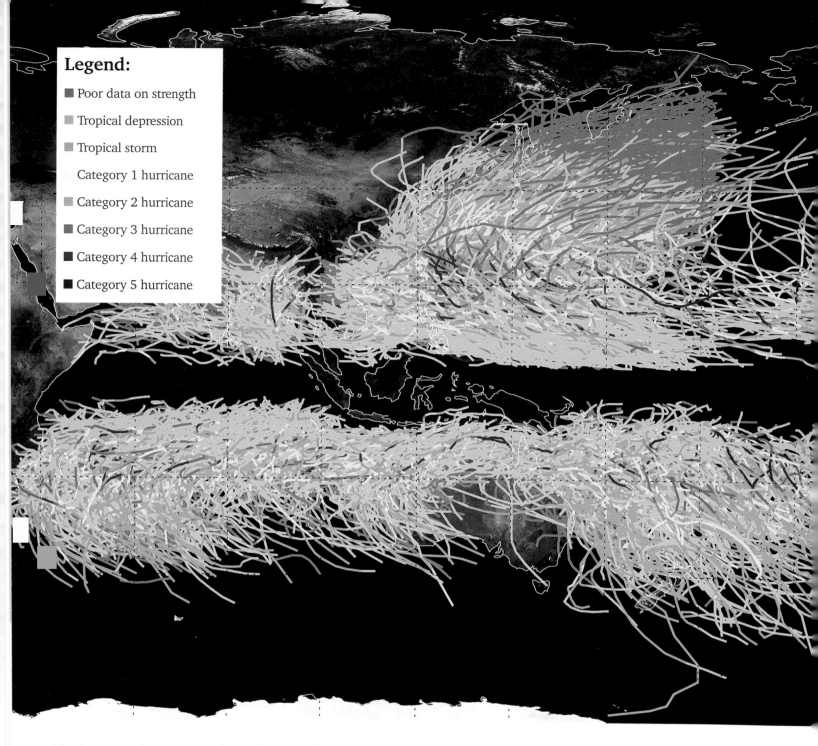

Legend:

- ■ Poor data on strength
- ■ Tropical depression
- ■ Tropical storm
- ☐ Category 1 hurricane
- ■ Category 2 hurricane
- ■ Category 3 hurricane
- ■ Category 4 hurricane
- ■ Category 5 hurricane

The image on these two pages, which was created by NOAA, shows all of the world's known tropical cyclones from 1947 through 2007.

amples. Jack Beven of the Hurricane Center points out that two other 2005 hurricanes strengthened "at a rate similar to the 1935 storm."

On September 21 and 22, 2005, Rita strengthened from a tropical storm with highest winds just shy of 74 mph (hurricane strength) to a Category 5 hurricane with winds of 167 mph in less than thirty-six hours. Fortunately, it weakened to barely a Category 3 storm with 115 mph winds before hitting extreme southwestern Louisiana just east of the Texas border.

Then, in the twenty-four hours ending at 1

a.m. on October 19 of the same year, Wilma strengthened from a 69 mph tropical storm to a 172 mph Category 5 hurricane, which the Hurricane Center described as "an unprecedented event for an Atlantic tropical cyclone. It is fortunate that this ultrarapid strengthening took place over open waters, apparently void of ships, and not just prior to a landfall." Wilma reached its peak sustained wind speed of 184 mph on October 19. Wilma weakened to a Category 4 hurricane with 150 mph winds before hitting the Mexican resort island of Cozumel on October 21. It weakened only slightly more be-

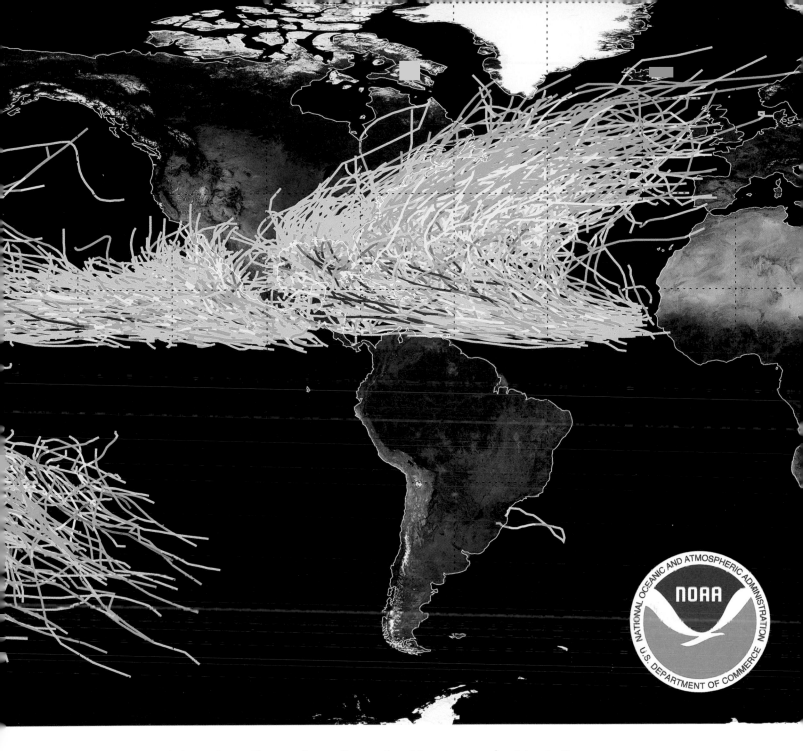

fore hitting Cancun six hours later. The two days between Wilma reaching Category 5 and its landfall gave Mexican authorities time to move thousands of residents and tourists to safety. In wreaking havoc on Cancun, Wilma could have killed many more than the four people who died in the storm.

We can only imagine how many people either of these hurricanes would have killed if they had rapidly intensified into a Category 5 storm less than two days before hitting places such as the Florida Keys, the Tampa Bay area, New Orleans, the Galveston-Houston area, or one of the many other heav-ily populated locations on the Atlantic Ocean or Gulf of Mexico coasts.

In this chapter and in the two preceding chapters, we've examined thunderstorms, tornadoes, organized groups of thunderstorms, weather fronts, and tropical cyclones that are responsible for most of the deadly events that from time to time make big weather news. In the next chapter, we will learn about a few kinds of less-dramatic weather that at times can be more deadly or costly than tornadoes or hurricanes.

Some weather phenomena you

rarely see in the news but that

can be destructive or deadly

CHAPTER 11

Early in the afternoon of August 3, 2006, NASA's Aura satellite, in its orbit 438 miles above the earth, measured air pollution in the atmosphere with its sensors as it passed over Washington, D.C.

At the same time, a score of scientists and students were working on a hazy afternoon at Howard University's Beltsville, Maryland, research site in temperatures approaching 100 degrees, to collect ground-truth data to help calibrate the satellite's measurements.

"The satellite is going around the world measuring pollution and water vapor," explained Belay B. Demoz, a NASA Administrator's Fellow at Howard University. "But how good are these measurements? The only way we know is by comparing—to see if the balloon and satellite measurements match."

For members of the Washington, D.C., press who had been invited to watch and talk with scientists, the afternoon's highlight was the preparation and launching of the weather balloon. It carried a regular NWS radiosonde (described in Chapter 6), an ozonesonde to measure ozone, and a device to measure humidity that's more precise than similar instruments on ordinary radiosondes. The launch was part of a NWS study of the accuracy of its radiosondes' humidity measurements.

Previous pages: Traffic and air pollution choke Mexico City in 2006. It's one of the many cities around the world coping with the dangers of unhealthy air. Such pollution often travels far, even crossing oceans.

Students from Howard and other universities were spending the summer working with the scientists, learning that research requires much hands-on labor as well as thinking in new ways about how the world works.

"I think it's good to have a wide range of abilities," said Bryan Ramson, a Howard University sophomore from New Orleans who's majoring in mathematics. Helping to launch weather balloons is "just one more thing I can say I know how to do. Maybe if I work for NASA, I may have to launch a balloon on Mars someday."

Bryan Ramson

"I'm the resident handy man," Ramson said. "If it's dangerous or heavy or needs to be fixed, I do it." When he was growing up, "I just started taking stuff apart because I wanted to fix it. In high school I took my computer apart because it was broken and I had no other way to fix it."

He explained that his biggest problem was writing programs in Fortran, the computer programming language developed in the 1950s for scientific computing that is still one of many being used today. "It was made to do certain things. When you try to make it do things it wasn't made to do, it screams

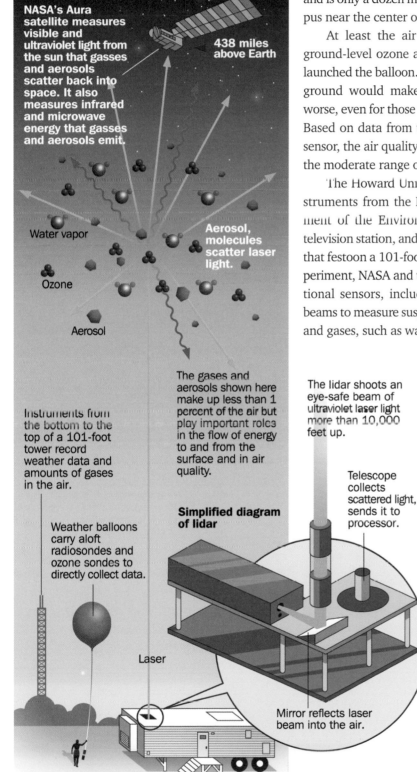

Air measurements from above and below

Scientists and students working at Howard University's Beltsville, Maryland, research site in 2006 and 2007 collected measurements of water vapor, ozone, other gasses, and aerosols in the air using a lidar (a laser radar), weather balloons, and instruments on the ground and on a tower. This information helped verify data from the Aura satellite.

NASA's Aura satellite measures visible and ultraviolet light from the sun that gasses and aerosols scatter back into space. It also measures infrared and microwave energy that gasses and aerosols emit.

438 miles above Earth

Water vapor

Ozone

Aerosol

Aerosol, molecules scatter laser light.

The gases and aerosols shown here make up less than 1 percent of the air but play important roles in the flow of energy to and from the surface and in air quality.

Instruments from the bottom to the top of a 101-foot tower record weather data and amounts of gases in the air.

Weather balloons carry aloft radiosondes and ozone sondes to directly collect data.

Laser

Simplified diagram of lidar

The lidar shoots an eye-safe beam of ultraviolet laser light more than 10,000 feet up.

Telescope collects scattered light, sends it to processor.

Mirror reflects laser beam into the air.

at you; it tells you it doesn't like that." Ramson hopes to work in quantum physics or for NASA after college.

Even though the Howard University site is an open field surrounded by trees, it's perfect for tracking the atmosphere and its pollution in an urban area. It's in the highly urbanized Interstate-95 corridor, which connects the East Coast's biggest cities, and is only a dozen miles from Howard's main campus near the center of Washington, D.C.

At least the air wasn't dangerously full of ground-level ozone as the scientists and students launched the balloon. High levels of ozone near the ground would make working in the heat even worse, even for those with no respiratory problems. Based on data from the site's ground-level ozone sensor, the air quality value in Beltsville was 79, in the moderate range on the air quality index.

The Howard University site is dotted with instruments from the NWS, the Maryland Department of the Environment, a Washington, D.C., television station, and NASA, including instruments that festoon a 101-foot tower. For the summer's experiment, NASA and three universities set up additional sensors, including lidars, which use laser beams to measure suspended, microscopic particles and gases, such as water vapor, in the air.

Instruments on the tower have to be maintained and that was fine with Lianna Samuel, a Howard University junior who's majoring in chemical engineering. She grew up spending a lot of time up in the fruit trees around her family's home in Trinidad and loves to climb. "I'm not afraid of heights; I could go up and work all day," she said. Before beginning work with the project, "I knew there were lots of instruments, but I didn't realize that we had towers that one [could] actually climb and connect instruments to."

Ozone near the ground is irritating and can be harmful to some people, as we see later in this chapter. High above the Earth, however, ozone blocks dangerous ultraviolet radiation, as we will see in Chapter 12.

Samuel was also writing computer programs to do such things as comparing the relations among measurements of humidity, temperature, ozone level, pressure, and height. Such results are helpful to scientists working on ozone prediction computer models as they provide data against which to test their models. She hopes to work as an environmental engineer after college.

Nyasha George, a Howard University senior physics major, also from Trinidad, said one aspect of the project she likes is using theory to answer practical questions, because "I like to be practical." The schedule can be hectic at times, she explained, because the ozonesonde launches are coordinated with the satellite as it passes overhead. "Last week over three days we must have had fifteen launches," which allowed little time for sleep, especially at night.

George, who started college as a joint dance-physics major, would like to work in dance medicine. She sees no conflict between science and the arts. In fact, she said, dance is "probably the highest way to apply physics because you need to be aware of physics within your body. It's deeper than just art or just dance."

While obtaining ground truth for satellite measurements is a big part of the project, the scientists and students are also learning more about the meteorology of ozone pollution in the area. "We have already observed examples of the influence of a narrow stream of strong winds during the night on surface-level ozone formation," said Howard's Everette Joseph, who leads the university's research team. The finding could help improve air quality forecasts.

The summer 2006 experiments were only the beginning, with similar programs during the winter of 2006–2007 and the summer and fall of 2007 to obtain measurements during different seasons.

"It's been very successful on a number of fronts. On pure scientific terms, we got a lot of good data for satellite validation, as well as also for studying local pollution and for studying local weather," said David Whiteman, who led the research team from NASA's Goddard Spaceflight Center in nearby Greenbelt, Maryland. Students and scientists collected data on polluted days and as cold fronts were coming through to push out the pollution. "It's important to do it when it's hot and polluted and clear and clean." Also, "it was an op-portunity for pulling together diverse researchers and students. That was pretty special and unique."

Demoz said, "Students are involved in leading-edge research," not just sitting in a classroom. "It has to be useful for them to get excited. This is how we can get them into meteorology; either it has to pay a lot, which it doesn't, or it has to be very exciting. By making it relevant, these guys get hooked."

Another advantage, he explained, is that students from Howard University, a historically black university founded in 1867; Pennsylvania State University; University of Colorado; Trinity University in Washington, D.C.; University of Wisconsin at Madison; and Smith College in Northampton, Massachusetts, all had to come in, sometimes at 2 a.m., to launch balloons. "They give each other rides, eat together, help each other; this breaks down a lot of biases."

The students also learned that nifty acronyms are a tradition in the atmospheric sciences. The campaign is called the Water Vapor Validation Experiment—Satellite/Sondes in 2006, which gives you "WAVES 2006."

Air pollution

As we saw in Chapter 6, meteorologists are looking for ways to capture images of weather "under the radar," such as the lower parts of tornado-producing supercell thunderstorms several miles from a weather office's radar antenna. In this chapter, we examine a few weather-related topics that are "under the radar" in the metaphorical sense—topics that you rarely see on the evening news, but which are important.

Air pollution, for example, rarely makes the headlines in the way that tornadoes, hurricanes, and floods do, but over time it can make many people ill, sometimes fatally ill. Other phenomena that share this characteristic, which we discuss in this chapter, are extremely hot or cold temperatures.

From early in the nineteenth century into the 1950s, black smoke pouring from factories was a sign of prosperity; it meant factory workers had jobs. Black smoke from the steam locomotives that pulled most of the world's trains meant they were busy hauling goods and materials, including the coal that heated homes, schools, offices, and other buildings in most industrialized nations.

Lianna Samuel

Nyasha George

Sharing science with students

When Belay Demoz earned his bachelor's degree in physics from Asmara University in Eritrea in 1984, he was thinking of becoming a nuclear physicist. But "we had a very bad drought, and I started thinking, 'What good is science if I can't do anything about this,'" he recalls. He had heard of cloud-seeding work being done in the United States and thought learning about it would be a way to help his East African nation. Demoz won an assistantship for graduate work at the University of Nevada–Reno, Desert Research Institute, where he focused on the inner workings of clouds, completing a doctoral thesis on Sierra Nevada winter storms, a study that used high-tech instruments.

Working on his master's and PhD degrees taught Demoz the benefits for students of working on real research. "The classroom teaching? You can just forget it after the exams. The hours of field-work tend to stay with you." Conducting research shows students what to expect when they graduate. Demoz recalls trips with John Hallet, one of his professors, to the countryside around Reno. Hallet "wanted us to learn and apply all of the mathematics. He would tell students, 'Look at that cloud and tell me what's going on.' Students would have to use theory they had learned in class, which is much more powerful than just memorizing."

After earning his PhD, Demoz conducted post doctoral research on cloud water at the University of Illinois. He was then a senior scientist at Hughes-STX and worked for the Joint Center for Earth Systems Technology (a collaboration between the University of Maryland, Baltimore County and NASA) before becoming a research meteorologist at NASA's Goddard Spaceflight Center in Greenbelt, Maryland, in 2000. Before coming to NASA, he helped Howard University win a NOAA grant to upgrade its Beltsville research site.

In 2005, NASA selected Demoz for a two-year term as a NASA Administrator's Fellow as part of a program that aims to enhance science, technology, engineering, and mathematics education at minority institutions. This enabled Demoz to teach for a year at Howard, while working for NASA. It also allowed him to work with Howard and other universities to organize and win NASA's approval of the WAVES experiment at the Beltsville site and to bring together students from around the United States and other nations for the experiment.

Demoz says working with students from different backgrounds "has real meaning for me. Coming from a small country, I wasn't privy to a lot of the cultural and social interactions you have in the United States. I see a lot of value here for us as a society to learn."

One of his hopes is to see fewer scientific meetings "where I'm the only black person." He is encouraging students from many backgrounds to participate in scientific meetings during his term as chair of the AMS Committee for Laser Atmospheric Studies, which ends in 2010.

His cloud physics studies showed Demoz that cloud seeding isn't the answer to East African drought. "Maybe it's not rain I can bring to Eritrea, but scientists. I think I'm the first meteorologist in Eritrea. Now there are six or seven who followed me. This has value although it didn't make it rain. Maybe the way I can help is by expanding to [the rest of] Africa; by bringing high-tech lidars and lasers and teaching students how to use them." He's convinced that working with "homegrown people" is the best way to address East Africa's climate issues.

Belay Demoz says since meteorology doesn't pay very much, it has to touch you. It has to move you.

Even though it meant the economy was booming, all of this coal smoke could be annoying and was often dangerous. While in the past some physicians had seen coal smoke as an antidote to the miasmas of night air, by the nineteenth century some were arguing that air pollution such as coal smoke was unhealthy even though they didn't have today's scientific evidence of exactly how some particles and gases in the air cause illnesses.

Any particle or gas in the air that could harm living things or damage property is considered to be **air pollution** if there is enough in the air at a particular time and place to cause harm. Air pollution can be natural or **anthropogenic** (caused by humans), but most of the debate about air pollution is about the anthropogenic particles and gases that humans add to the air.

Natural sources of pollutants include dust, smoke from wildfires, volcanic eruptions, and even pine and similar trees that emit the **hydrocarbons** known as **volatile organic compounds (VOCs)**, which we notice as a fresh, pine-tree smell. VOCs from trees help form a bluish haze, which gives the Blue Ridge Mountains of Virginia and the Blue Mountains of Australia their names. Unfortunately, haze created by hydrocarbons and other pollutants that humans add to the air is making natural haze thicker in many places.

Attempts to control air pollution go back to at least 1306 when King Edward I of England "issued a proclamation banning the use of sea coal in London due to the smoke it caused," James R. Fleming and Bethany R. Knorr write in their "History of the Clean Air Act" on the American Meteorological Society Web site.

Volatile organic compounds include a variety of chemicals emitted by thousands of products, such as paints and lacquers; cleaning supplies; pesticides; building materials and furnishings; and graphics and craft materials, such as glues and adhesives, permanent markers, and photographic solutions. These compounds are a major source of indoor air pollution, which can be worse in some cases than outdoor pollution. Natural sources of volatile organic compounds include plants and plankton in the ocean.

During the nineteenth century, London became notorious for its thick, polluted fog. In some specific cases, thick pollution killed scores or even hundreds of people. In 1905, physician Henry An-

toine Des Voeux coined the term "smog" to mean a combination of smoke and fog. He told a London newspaper that it "required no science to see that there was something produced in great cities which was not found in the country, and that was smoky fog, or what was known as smog."

Limiting air pollution. Fleming and Knorr note that when U.S. cities, including Chicago and Cincinnati, adopted clean air laws in 1881, other cities followed in the coming years. However, no effective federal action was taken until the 1950s.

Many studies over several years have found correlations between different kinds of air pollution and deaths, but determining that a particular person died of "air pollution" is difficult. Air pollution's effects range from irritation of the eyes and lungs, tightness in the chest, and headaches to respiratory illnesses, chronic lung diseases such as asthma and emphysema, and aggravation of heart disease. Those who die of these effects are usually counted as air pollution victims only in statistical studies linking amounts of air pollution to various causes of illness and deaths.

Today, the only time you'll see an air pollution story on television with yellow police-line tape, flashing red and blue lights, and emergency medical technicians tending to victims is when a toxic substance has been released. If today's television news had been around in October 1948, however, you would have seen firefighters carrying oxygen tanks rushing to aid victims of an air pollution disaster caused by ordinary emissions from factories affecting the entire town of Donora, Pennsylvania.

Response to a disaster. On Friday evening, October 28, 1948, thick gray fog covered the town of Donora in the Monongahela River Valley, 20 miles southeast of Pittsburgh. But "no one paid much attention to it," Jeff Gammage wrote in the *Philadelphia Inquirer* on October 18, 1998, the disaster's fiftieth anniversary. "The people here were long accustomed to the occasional gassy cloud that spewed from the local steel plant. "By Saturday night, 11 people were dead, choked to death by the noxious cloud. Nine more died in the ensuing hours. By Monday, nearly 7,000 people—half the town's population—were ill at home or in hospitals, sickened by a lethal mix of sulfur dioxide, carbon monoxide, and metal dust."

"Sea coal" is found near the surface of the ground. The English call it "sea coal" because many deposits are found near the English coast. It produces more noxious gases when burned than other kinds of coal.

Bill Schempp, 81, a volunteer fireman who toted tanks of oxygen to the stricken, told Gammage, "There wasn't a damn thing you could do about it. We had stagnant, noxious air, and it wasn't moving." Gammage summed up residents' attitudes in 1948: "People figured the sporadic smelly fog that drifted from the mill was the price they paid for an abundance of good jobs."

The Donora disaster and serious air quality problems in Los Angeles and other cities "raised public awareness and concern about this issue once again," Fleming and Knorr wrote. "In 1955, the government decided that this problem needed to be dealt with on a national level. The Air Pollution Control Act of 1955 was the first in a series of clean air and air quality control acts, which are still in effect and continue to be revised and amended."

Weather and air pollution. The substances that killed and injured people in Donora were the same pollutants that factories in and around the town had been emitting for years. The disaster occurred because the weather kept winds from carrying the pollutants away and diluting the toxins as they mixed with clean air.

An inversion—air aloft warmer than ground-level air—makes the atmosphere very stable and acts like a lid over a region. Local inversions can form overnight when the sky is clear and heat easily radiates away into space. This rapidly cools the air near the ground while air aloft cools more slowly. If thick clouds don't block solar radiation, this kind of inversion will disappear the next day as the sun heats the ground.

There was a **subsidence inversion** over Donora caused by air descending and warming in a large area of high atmospheric pressure. Air that's descending in such an area normally doesn't make it all of the way to the ground, but only to the top of what's known as the **mixing layer**. Air in the mixing layer rises and cools until it reaches the top of the layer, which is where the rising air is no longer warmer than the surrounding air.

In the Donora event, a weak storm arrived after five days, replacing the high pressure that had been dominating the Northeast. The storm's winds diluted and pushed away the deadly pollution.

Dangerous small particles. Air pollution includes a variety of gases and aerosols, which are

Clear above, murky below

An air traveler through the 1940s might have seen a view like the one below while flying above any industrial town in the United States. While the contrast today between upper and lower levels of the sky isn't as great as before clean air laws were adopted, a low-level layer of pollution is still seen when conditions are like those shown here.

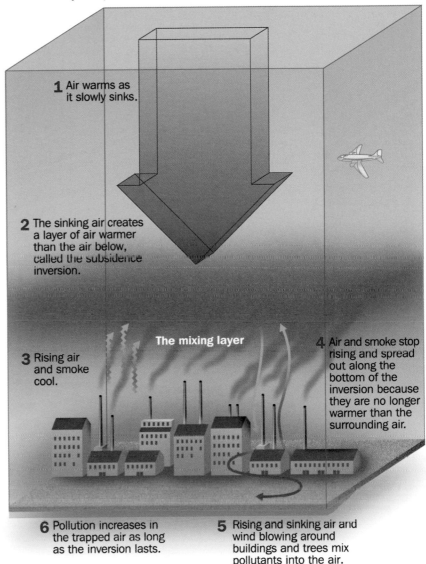

1 Air warms as it slowly sinks.

2 The sinking air creates a layer of air warmer than the air below, called the subsidence inversion.

The mixing layer

3 Rising air and smoke cool.

4 Air and smoke stop rising and spread out along the bottom of the inversion because they are no longer warmer than the surrounding air.

6 Pollution increases in the trapped air as long as the inversion lasts.

5 Rising and sinking air and wind blowing around buildings and trees mix pollutants into the air.

liquids or solid particles floating in the air. While larger solid particles can block visibility and make life uncomfortable, the ones that cause health concerns are smaller than ten micrometers in diameter because these can be inhaled.

Particles smaller than two-and-a-half micrometers are especially dangerous because they can travel deep into your lungs, even down to the alveoli. This is where the oxygen you breathe goes into your blood and the carbon dioxide your body produces goes into the air to be exhaled.

Smoke, such as from wildfires, contains these tiny particles. Chemical reactions among gases

Dangerous tiny particles

Very tiny particles are one of the five major air pollutants the U.S. Environmental Protection Agency uses to calculate air quality. Burning of fossil fuels, such as gasoline and coal, and other kinds of combustion, such as wood burning in stoves or forest fires, are major sources of tiny particles. Some industrial processes can also add them to the air.

Relative sizes of particles

2.5-micrometer particle in relation to a 10-micrometer particle. Small enough to pass deep into lungs.

10-micrometer particle in relation to a hair. Small enough to pass through nose and throat into lungs.

Diameter of human hair approximately 70 micrometers.

Fine beach sand 90 micrometers.

Effects of particles

Particles smaller than 10 micrometers cause inflammation and other problems in the bronchi (airways).

Alveoli enlarged

Particles smaller than 2.5 micrometers become trapped, aggravating bronchitis, emphysema, and heart disease.

VOG—from the words "volcano" and "fog" —is used in Hawaii for a dangerous mixture of gases that includes tiny particles and droplets formed when gases from the Kilauea Volcano react with moisture, oxygen, and sunlight.

emitted from motor vehicles, power plants, and factories can also form them.

Natural air cleaning. We note above that wind ends pollution episodes, like the one in Donora, by blowing the pollutants away and mixing them with cleaner air. In fact, pollution didn't kill scores of people every week in Donora and other similar industrial towns and cities because most of the time convection, the air's vertical motions due to surface heating, and ordinary winds diluted the pollution to levels that weren't lethal or even noticeable.

If nothing else happened, however—if the pollution stayed in the air forever—levels would ultimately increase all around the earth. Fortunately for life on Earth, the air cleans itself, although humans have managed to upset the natural balance for some pollutants in many places.

Clouds and precipitation are nature's most important way of cleaning the air. As we saw in Chapter 4, cloud droplets form on condensation nuclei, which can include pollution aerosols. When rain or snow fall they sweep more aerosols from the sky, which is why the air often smells fresh after a rainstorm.

Of course the pollutants that help form water droplets and are swept from the sky by falling precipitation end up on the ground. During the days when black smoke filled the skies of industrial areas, rain would coat everything it hit with soot. That doesn't happen in industrialized nations today, but what does happen is worse in some ways because it's subtle.

Acid rain. By the 1960s, hikers and campers who were familiar with wilderness areas in the Northeast, such as New York State's Adirondack Mountains, began to notice that the water in some of the many ponds in the mountains was clearer than before—a sign the water had lost much of its near-microscopic-size life. They heard fewer frogs at night because the ponds no longer supported the insects and other small creatures the frogs ate.

At the time, a few scientists were beginning to talk about **acid rain** damaging the environment of the Adirondacks and other areas. Parts of Europe, including Germany's famous Black Forest, were experiencing similar damage. Acid rain also damages buildings made of limestone and marble, making the surfaces rough, removing material, and blurring carved details of statues.

For the northeastern United States, the problem began with coal-burning power plants and other sources in the Midwest that emitted sulfur and nitrogen compounds, which the winds carried to the east. These compounds dissolved in raindrops and snow crystals to form sulfuric acid and carbonic acid, which fell on places hundreds of miles away from the pollution sources. Acid rain most affects locations such as New York's Adirondack Mountains, which have soils that don't naturally neutralize acids, as soils in many areas do.

Today's cleaner air. Technological and social changes, such as the replacement of coal with natural gas as the major heating fuel along with clean air laws and regulations in many parts of the world, have eliminated most of the obvious pollution that darkened skies into the 1950s, at least in many wealthy nations, but such pollution is becoming a big concern in nations such as China and India.

No industrialized nation is likely to see an event like the thick, smelly smog that killed an estimated 4,000 people in London between December 5 through 9, 1952. Instead of the sulfur-laden London smog of 1952, which was produced in large part by burning fuel oil and coal for heat and factory production, today residents of the United States and Europe worry about **photochemical smog**. This is created by chemical changes in the atmosphere involving nitrogen oxides and hydrocarbons, including volatile organic compounds, that create aerosols and also ozone.

The Clean Air Act. The 1990 U.S. Clean Air Act sets standards for maximum amounts of various kinds of air pollution. All states have to meet the national standards but states that wish to can set stricter standards. The law also requires states to submit plans for cleaning up pollution, which the U.S. Environmental Protection Agency (EPA) must approve.

Under the Act, the EPA sets limits for the amounts of carbon monoxide, lead, nitrogen dioxide, ozone, particulate matter (tiny particles), and sulfur dioxide in the air. A variety of sources directly emit carbon monoxide, lead, nitrogen dioxide, and sulfur dioxide. Some sources directly emit particulate matter. Chemical reactions in the air powered

The Air Quality Index

The Air Quality Index (AQI) is based on measurements of ground-level ozone, particle pollution, carbon dioxide, sulfur dioxide, and nitrogen dioxide. U.S. Environmental Protection Agency formulas are used to calculate a value for each pollutant. The highest value is that day's AQI. For example, if an area had AQI values of 90 for ozone and 88 for sulfur dioxide, the AQI value would be 90 for the pollutant ozone that day.

Air quality	AQI numbers	Meaning
Good	0-50	Satisfactory. Little or no risk to anyone
Moderate	51-100	Acceptable. Moderate health concern for unusually sensitive people
Unhealthy for sensitive people	101-150	Members of sensitive groups may experience health effects
Unhealthy	151-200	Everyone may begin to experience health effects; more serious for sensitive groups
Very unhealthy	201-300	Health alert. Everyone may experience more serious health effects
Hazardous	More than 300	Emergency conditions. The entire population is more likely to be affected.

by the sun and involving nitrogen oxides and volatile organic compounds form dangerous tiny particles and also ozone.

The Air Quality Index. The EPA has developed formulas to calculate the degree of danger of air pollution, using numbers to represent levels of either ozone or particle concentrations and tying them to the levels of air pollution danger.

Air quality levels can be based on the actual observations of pollutants in the air or NWS forecasts of pollution levels expected to occur for the day the forecast is made and the following day. To make these forecasts, the Weather Service combines its most advanced operational computer forecasting model and a model to forecast how the weather will transport various chemicals.

Good and bad ozone. Ozone is a form of oxygen with each molecule consisting of three oxygen atoms (each molecule of the more common form of oxygen that makes up approximately 21 percent of the atmosphere has two oxygen atoms). Whether ozone is harmful or helpful depends on its location. Ozone near the ground can make your eyes sting and irritate your respiratory system, which is especially troublesome for those suffering from asthma or bronchitis, although it can be uncomfortable for healthy people as well. It also harms plants and is estimated to do billions of

Air pollution

Photochemical smog is the most common kind of air pollution, accounting for nearly 70 percent of the days with "unhealthy" air quality in the United States from 2004 through 2006, according to the U.S. Environmental Protection Agency. Natural substances play a role, but emissions created by burning fuel, such as in vehicles and power plants, are needed to create most smog.

A smoggy day

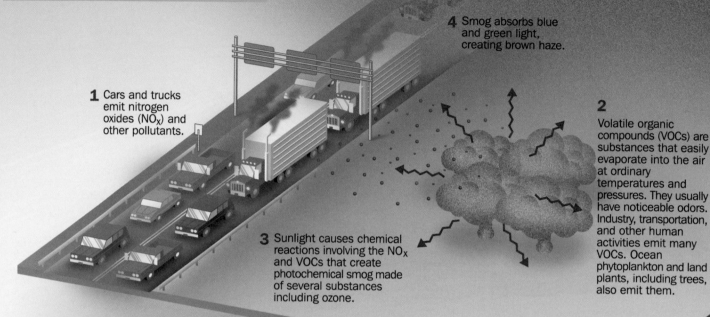

4 Smog absorbs blue and green light, creating brown haze.

1 Cars and trucks emit nitrogen oxides (NO_x) and other pollutants.

2 Volatile organic compounds (VOCs) are substances that easily evaporate into the air at ordinary temperatures and pressures. They usually have noticeable odors. Industry, transportation, and other human activities emit many VOCs. Ocean phytoplankton and land plants, including trees, also emit them.

3 Sunlight causes chemical reactions involving the NO_x and VOCs that create photochemical smog made of several substances including ozone.

Air pollution and climate

Air pollution has traditionally been considered a regional or local problem; it can be linked with regional and even global climate. Pollution aerosols can warm the upper atmosphere and cool the surface, which affects atmospheric pressures and thus weather patterns, such as the Asian monsoon. They can also affect precipitation.

Many tiny water drops condense on aerosols in clouds. These reflect away more solar energy than the fewer larger drops that otherwise would have formed from the same amount of water vapor.

Aerosols such as black carbon, dust, sulfates, nitrates, and ash absorb solar energy, which warms the air aloft and cools the surface.

Warming the air aloft makes the atmosphere more stable, reducing convection that could lead to rain.

Absorption and scattering reduces solar energy reaching the surface.

Smaller cloud drops are less likely to grow to raindrop size than larger drops.

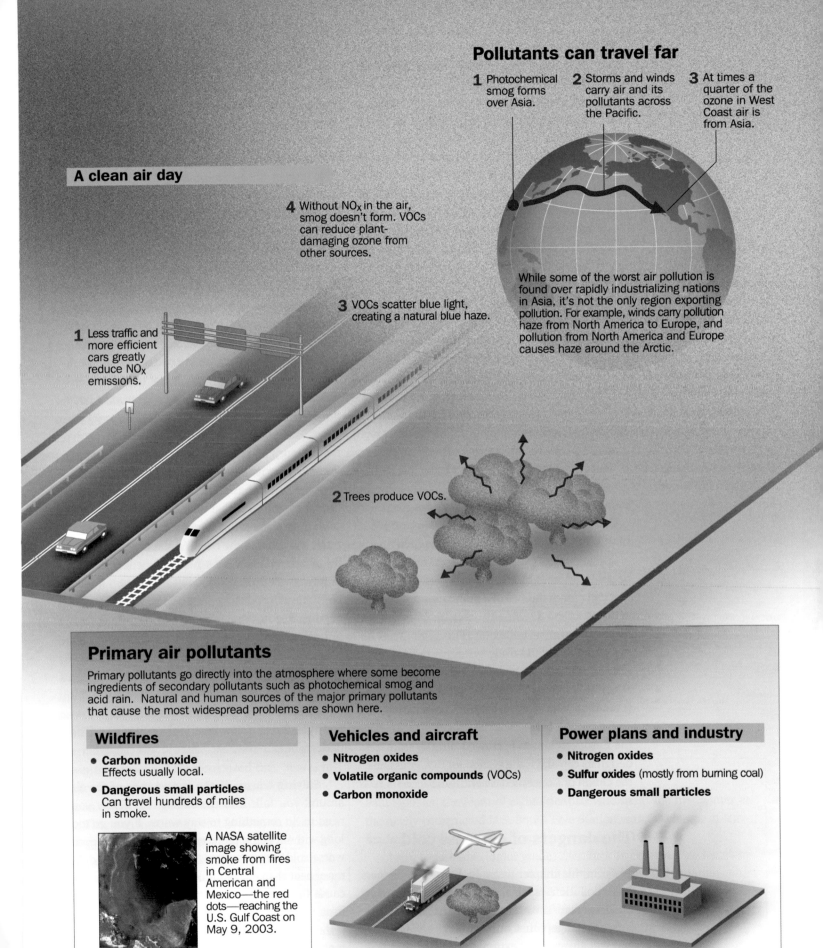

Pollutants can travel far

1 Photochemical smog forms over Asia.

2 Storms and winds carry air and its pollutants across the Pacific.

3 At times a quarter of the ozone in West Coast air is from Asia.

While some of the worst air pollution is found over rapidly industrializing nations in Asia, it's not the only region exporting pollution. For example, winds carry pollution haze from North America to Europe, and pollution from North America and Europe causes haze around the Arctic.

A clean air day

4 Without NO$_x$ in the air, smog doesn't form. VOCs can reduce plant-damaging ozone from other sources.

3 VOCs scatter blue light, creating a natural blue haze.

1 Less traffic and more efficient cars greatly reduce NO$_x$ emissions.

2 Trees produce VOCs.

Primary air pollutants

Primary pollutants go directly into the atmosphere where some become ingredients of secondary pollutants such as photochemical smog and acid rain. Natural and human sources of the major primary pollutants that cause the most widespread problems are shown here.

Wildfires

- **Carbon monoxide** Effects usually local.
- **Dangerous small particles** Can travel hundreds of miles in smoke.

A NASA satellite image showing smoke from fires in Central American and Mexico—the red dots—reaching the U.S. Gulf Coast on May 9, 2003.

Vehicles and aircraft

- **Nitrogen oxides**
- **Volatile organic compounds** (VOCs)
- **Carbon monoxide**

Power plans and industry

- **Nitrogen oxides**
- **Sulfur oxides** (mostly from burning coal)
- **Dangerous small particles**

Carbon dioxide has not traditionally been considered an air pollutant, but in April 2007 the U.S. Supreme Court ruled that it and other greenhouse gases are "air pollutants" under the U.S. Clean Air Act.

Wind chill chart

This is the wind chill chart that the U.S. NWS adopted in 2001. The actual numbers aren't as important as the colors showing how long frostbite of skin exposed to the cold and wind could take to occur. Any wind chill chart is at best a rough guide to the dangers of cold and wind because it doesn't include individual differences or factors such as sunshine. For example, the NWS says bright sunshine increases wind chill temperatures by 10 to 18 Fahrenheit degrees.

Frostbite Times   ▨ 30 minutes   ▨ 10 minutes   ■ 5 minutes

Wind (mph)	\ Temperature (°F)	40	35	30	25	20	15	10	5	0	-5	-10	-15	-20	-25	-30	-35	-40	-45
Calm																			
5		36	31	25	19	13	7	1	-5	-11	-16	-22	-28	-34	-40	-46	-52	-57	-63
10		34	27	21	15	9	3	-4	-10	-16	-22	-28	-35	-41	-47	-53	-59	-66	-72
15		32	25	19	13	6	0	-7	-13	-19	-26	-32	-39	-45	-51	-58	-64	-71	-77
20		30	24	17	11	4	-2	-9	-15	-22	-29	-35	-42	-48	-55	-61	-68	-74	-81
25		29	23	16	9	3	-4	-11	-17	-24	-31	-37	-44	-51	-58	-64	-71	-78	-84
30		28	22	15	8	1	-5	-12	-19	-26	-33	-39	-46	-53	-60	-67	-73	-80	-87
35		28	21	14	7	0	-7	-14	-21	-27	-34	-41	-48	-55	-62	-69	-76	-82	-89
40		27	20	13	6	-1	-8	-15	-22	-29	-36	-43	-50	-57	-64	-71	-78	-84	-91
45		26	19	12	5	-2	-9	-16	-23	-30	-37	-44	-51	-58	-65	-72	-79	-86	-93
50		26	19	12	4	-3	-10	-17	-24	-31	-38	-45	-52	-60	-67	-74	-81	-88	-95
55		25	18	11	4	-3	-11	-18	-25	-32	-39	-46	-54	-61	-68	-75	-82	-89	-97
60		25	17	10	3	-4	-11	-19	-26	-33	-40	-48	-55	-62	-69	-76	-84	-91	-98

at Canada's Defence Civil Institute of Environmental Medicine in Toronto. This led to the chart and formulas used today. The figures produced by these new formulas aren't as extreme as those of the old one. For example, for a 5 degree temperature and 30 mph wind, the old formula gave a wind chill of −40 degrees, while the new formula yields −19 degrees.

The dangers of sunburn

Worrying about the wind chill as you prepare to leave home for work can lead to dreams of catching the next flight to a place that's sunny and warm. If you catch that flight, you should remember that sun and warmth have their own dangers, as we will see below.

Most of us think of sunburn as a warm-weather danger, but it's also a concern when you're surrounded by ice and snow. In fact, if you're skiing in Colorado or working in Antarctica, you need to worry about sunburn because ice and snow reflect the ultraviolet energy that causes sunburn even though the sun isn't nearly as high in the sky as in the tropics or in the summer.

In 1994 the National Weather Service and the Environmental Protection Agency developed a UV (ultraviolet) Index to characterize the danger of sunburn, and the NWS began forecasting the index

value. In 2004 the NWS and the EPA revised the index to make it consistent with World Health Organization and World Meteorological Organization guidelines. Sunburn's danger goes far beyond the pain it causes in the hours after it occurs. Medical researchers have found strong links between sunburn and skin cancer.

The dangers of extreme heat

When the air's heat begins to warm your body above approximately 99 degrees, blood vessels dilate to allow more blood to flow through them, which means more blood circulates through the tiny capillaries right under the outer layer of your skin, and your heart has to pump harder. It's easy to see why hot weather is especially dangerous to those with heart or circulation ills.

As the added blood flow warms your skin, more heat radiates into the air. At the same time, your sweat glands begin secreting water. Your body supplies the heat needed for this perspiration to evaporate into the air. That is, some of your body's heat moves into the air in the form of latent heat in your evaporated perspiration.

High humidity, however, slows evaporation. In such conditions, this means perspiration makes you sticky instead of cooling you. The NWS uses its **heat index** chart to calculate an **apparent temperature**. While the humidity doesn't change the air's temperature, a high level has the same effect as a higher temperature by inhibiting evaporative cooling. The heat index chart lists danger zones for different apparent temperatures, but these are very general. The "extreme caution" zone for one person could be the "extreme danger" zone for someone else. The best general rule is to go easy on outdoor exercise and drink plenty of water when the heat index moves into the "caution" zone. Your body can become more tolerant of heat, but you should build up this tolerance slowly.

When you become hotter and perspire more, the problems likely to arise first result from a loss of water and an imbalance of salts (since your body loses water and salt when you perspire). Thus, the advice to "stay hydrated." Some of the first symptoms of overheating include cramps in the leg or abdomen muscles. Lying down in a cool place and massaging cramped muscles, while drinking water

The UV index and sunburn danger

The U.S. NWS and other weather services around the world forecast a daily UV (ultraviolet) Index as a guide to the danger of sunburn. In general, the danger of sunburn increases with the height of the sun in the sky, which means the danger is always greater in the tropics than in higher latitudes.

Puffy clouds reflect UV energy, making total higher than from direct sunlight alone.

Thin clouds allow most UV energy to reach the ground.

Shade doesn't provide 100% sunburn protection.

Beach sand and snow scatter UV energy.

The sky scatters UV energy just as it scatters blue light

UV rays are 80% as strong under a foot of water as in the air.

The UV Index

UV index number	Exposure Level
2 or less	Low
3–5	Moderate
6–7	High
8–10	Very high
11+	Extreme

How the UV index changes during the day

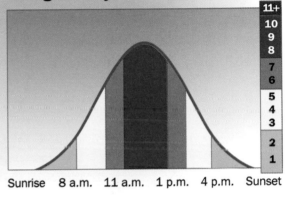

Sunrise   8 a.m.   11 a.m.   1 p.m.   4 p.m.   Sunset

(avoid alcoholic and sugary beverages) should be the first thing you try and could be all you need to do as long as you take it easy.

Heat exhaustion, which is characterized by heavy sweating, weakness, pale and clammy skin, and possible fainting or vomiting, is more serious. First aid should include getting the victim out of the sun, laying the victim down, and making him or her comfortable (loosen clothing); an air-conditioned place is best. The victim should also sip water.

Heat stroke, which is also called sunstroke, is a medical emergency and if it occurs you should call an ambulance or quickly get the victim to a hospital. A higher body temperature, hot and dry skin, and a rapid and strong pulse are symptoms of heat stroke. The victim could be unconscious.

An underrated killer. Most people tend to think of hot weather as uncomfortable and a pos-

sible danger for those who insist on going out for a long run when the heat index is in the danger zone. They aren't likely to think of heat as being as deadly as a hurricane, tornado, or blizzard. In Chapter 1, we discussed the fuzzy nature of statistics for weather-related deaths in the United States. Though fuzzy in some respects, they definitely show that heat is one of the leading causes of weather-related deaths—a bigger killer than tornadoes, for example.

The scene at the Cook County, Illinois, Medical Examiners Office in Chicago on July 15, 1995, during the city's worst weather disaster, a heat wave, illustrates how underrated a killer a heat wave can be. "Hundreds of victims never made it to a hospital. The most overcrowded place in the city was the Cook County Medical Examiners Office, where police transported hundreds of bodies for autopsies," Eric Klinenberg says in an interview published on the University of Chicago Press Web site regarding

The heat index

The heat index is a general guide to how the danger of heat increases as the temperature and relative humidity increase. Effects on individuals will vary greatly. The figures are based on effects of prolonged exposure when you are in the shade. Direct sunlight can increase index values by as much as 15 degrees Fahrenheit.

Relative Humidity (in percent)

Air Temp (F°)	0	5	10	15	20	25	30	35	40	45	50	55	60	65	70	75	80	85	90	95	100
140	125																				
135	120	128																			
130	117	122	131																		
125	111	116	123	131	141																
120	107	111	116	123	130	139	148														
115	103	107	111	115	120	127	135	143	151												
110	99	102	105	108	112	117	123	130	137	143	150										
105	95	97	100	102	105	109	113	118	123	129	135	142	149								
100	91	93	95	97	99	101	104	107	110	115	120	126	132	138	144						
95	87	88	90	91	93	94	96	98	101	104	107	110	114	119	124	130	136				
90	83	84	85	86	87	88	90	91	93	95	96	98	100	102	106	109	113	117	122		
85	78	79	80	81	82	83	84	85	86	87	88	89	90	91	93	95	97	99	102	105	108
80	73	74	75	76	77	77	78	79	79	80	81	81	82	83	85	86	86	87	88	89	91
75	69	69	70	71	72	72	73	73	74	74	75	75	76	76	77	77	78	78	79	79	80
70	64	64	65	65	66	66	67	67	68	68	69	69	70	70	70	70	71	71	71	71	72

80°–90° Caution: Fatigue

90°–104° Extreme Caution: Sunstroke, heat cramps, heat exhaustion possible

105°–129° Danger: Sunstroke, heat cramps, heat exhaustion likely; heatstroke possible

130° or higher Extreme Danger: Heatstroke likely except for brief exposure

After many weather-related disasters, NOAA appointed a committee to examine how the NWS and other organizations handled each disaster, with the goal of discovering what went wrong and suggesting ways to avoid similar mistakes.

his 2003 book *Heat Wave: A Social Autopsy of Disaster in Chicago*. "The morgue typically receives about 17 bodies a day and has a total of 222 bays …Just three days into the heat wave…its capacity was exceeded by hundreds, and the county had to bring in a fleet of refrigerated trucks to store the bodies. Police officers had to wait as long as three hours for a worker to receive the body. It was gruesome and incredible for this to be happening in the middle of a modern American city."

Heat "may be one of the most underrated and least understood of the deadly weather phenomena," NOAA said in its Natural Disaster Survey Report on the 1995 Midwest heat wave. "In contrast to the visible, destructive, and violent nature associated with 'deadly weather,' like floods, hurricanes, and tornadoes, a heat wave is a 'silent disaster.' Unlike violent weather events that cause extensive physical destruction and whose victims are easily discernible, the hazards of extreme heat are dramatically less apparent, especially at the onset."

NOAA's July *1995 Heat Wave* report notes that not only were the afternoons hot, but overnight temperatures fell only into the eighties and the humidity was high. The high humidity and high overnight temperatures are important because those who don't have air-conditioned places to sleep have no respite from the heat.

Heat waves and the poor. High overnight temperatures like those in the Midwest in 1995 mean that those without air conditioning find no relief. As in other cities at other times, those killed by the Midwest heat were disproportionally old and poor. Klinenberg says, "Hundreds of Chicago residents died alone, behind locked doors and sealed windows, out of contact with friends, family, and neighbors, unassisted by public agencies or community groups. There's nothing natural about that." The elderly and ill are especially vulnerable to heat stress because age and some medications can weaken the body's heat defenses.

Heat waves during the 1990s taught the NWS and public officials in cities across the United States that the threat of a heat wave should prompt public action just as the threat of a hurricane does. Actions include opening cooling centers where individuals who can't afford air conditioning can go for at least a few hours of relief from the heat and ensuring that elderly or ill men and women who live by themselves aren't left to die simply because no one is checking on them.

European nations began learning similar lessons in 2003 when a July and August heat wave, which broke records in several nations for the highest temperatures ever recorded, killed tens of thousands. France, the worst affected, estimated that nearly 15,000 people, mostly elderly, died from heat that summer. This was roughly half of the total estimated death toll from heat across Europe that summer.

Deadly heat waves in the United States and in Europe show that residents of cities that rarely experience temperatures in the high nineties suffer the most. This reflects the fact that in cities such as Chicago or Paris, staying warm in winter is a bigger concern than staying cool in the summer; apartment buildings and homes built to be warm in winter can become heat traps during the summer, especially if the residents are too poor to afford air conditioning.

As many American tourists have learned over the years, air conditioning is not nearly as common in Europe as in the United States. Even a city such as Chicago, which averages only seventeen days a year with temperatures warming above ninety degrees, has more homes, hotels, businesses, restaurants, and other public places with air conditioning than European cities such as Paris, London, or Berlin. This makes sense, however, as Chicago has warmer summers with an average high July temperature of 84 degrees. In Europe the average July highs are: London, 71; Berlin, 73; and Paris, 75. Also, electricity costs more in Europe, which means air conditioning isn't as affordable.

Looking to the future. A European Union (EU) report on the 2003 heat wave stated that finding ways to address the growing risk of heat deaths should be a priority of the EU, because global warming is expected to make heat waves more common while Europe's aging population means

the proportion of people over sixty and more vulnerable to heat is growing.

For instance, a 2002 report by the United Nations Population Division found that in Italy, one of the European nations that is aging the quickest, 24 percent of the population was older than sixty in 2000 and forecasts that the share of Italians who are sixty or older will grow to 34 percent in 2025 and 42 percent in 2050.

The United Nations report found that while the population of the United States isn't aging as quickly as most European nations, it is growing older. In the United States, 16 percent of the population was sixty or older in 2000. The United Nations forecasts that by 2025, the U.S. population sixty and older will increase to 25 percent, and by 2050 to 27 percent. Obviously, finding ways to reduce the vulnerability of an aging population to heat should be a concern in the United States.

Summary and looking ahead

The Howard University students we met in the opening of this chapter were working with scientists to better understand air pollution—an atmospheric phenomenon that is likely to affect more people around the world in the coming years than more high-profile disasters such as hurricanes and tornadoes. Even those who live in places such as parts of the United States west of the Rockies or in Europe, where hurricanes never hit and tornadoes are rare, can't escape air pollution from far-away places.

And both frigid and hot weather will always be dangerous for those who don't know how to cope, or who can't afford to cope.

The Howard University students will also be living in a world with a changing climate. In Chapter 12, we will examine possible effects of climate change associated with an average warming of the earth among the many weather and climate challenges facing everyone in the coming years.

That November representatives of thirty-five nations, including the United States, met in Geneva, Switzerland, to set up the Intergovernmental Panel on Climate Change (IPCC) under the United Nations Environmental Programme and the World Meteorological Organization. Its tasks were to:

- Assess the scientific evidence for a global warming trend and, if such a trend had started, look into its causes.
- Analyze the impacts of climate change.
- Formulate possible responses.

The IPCC was much different from previous groups concerned with climate, as Spencer Weart points out on the Web site for his book, *The Discovery of Global Warming*. "Unlike earlier conferences, national academy panels, and advisory committees, the IPCC was composed mainly of people who participated not only as science experts but as official representatives of their governments…The IPCC was neither a strictly scientific nor a strictly political body, but a unique hybrid."

The IPCC does not conduct research on its own but uses committees of scientists to evaluate research on various aspects of the climate from around the world. Three working groups prepare the IPCC's reports, with the 2007 series of reports being the fourth since they first came out in 1990.

The IPCC says that 450 lead authors from 130 nations wrote the 2007 reports. Their work was based on contributions by more than 800 contributing authors and 2,500 experts who reviewed parts of the reports.

The report from the IPCC's Working Group I on the scientific basis usually receives the most attention, because it represents the considered opinion of climate scientists on what is known about climate change and what needs to be learned. The report from Working Group II describes impacts, adaptations, and vulnerabilities, and the one from Working Group III covers mitigation and policy options. The IPCC also prepares a synthesis report summing up the three reports.

The first IPCC report, which came out in 1990, said the world had warmed during the twentieth century, but the science of the time couldn't sort out how much of this was natural and how much was caused by the carbon dioxide and other greenhouse gases that humans were adding to the atmosphere.

In its 1995 report, after noting uncertainties in the science, the IPCC concluded that the world was getting warmer and said, "The balance of evidence suggests that there is a discernible human influence on global climate." In its next report, in 2001, the Working Group I panel noted that both humans and nature can cause climate change, but indicated, "Most [66%] of the warming of the past fifty years is likely to be attributable to human activities."

The 2007 IPCC report. In its report *Climate Change 2007: The Physical Science Basis*, the IPCC Working Group I wrote in the summary for policymakers that scientific understanding of both warming and cooling influences caused by humans on global climate led its members to have "very high confidence" that human activities since 1750 had caused the global climate to warm. "Very high confidence" means "at least a nine out of ten chance of being correct," the report says.

The report gives ranges for potential warming and sea level rise based on six scenarios for how much carbon dioxide and other greenhouse gases are added to the atmosphere by the end of the twenty-first century. These take into account factors such as population and economic growth plus potential technological advances that could increase or decrease the rate at which humans are adding carbon dioxide to the air.

The amount of a gas such as carbon dioxide (CO_2) in the air can be measured in **parts per million (ppm)**, which is the ratio of the number of molecules of the gas to the total number of dry-air molecules. For example, 280 ppm CO_2 is 280 CO_2 molecules for each million molecules of dry air. Ice cores and other paleoclimatological data show that in 1750, before the Industrial Revolution began, the atmosphere contained 280 ppm of carbon dioxide. Direct measurements show that by 2005 this had increased to 379 ppm. In addition, the amounts of other greenhouse gases, such as methane and nitrous oxide, have also increased since pre-industrial times. In its 2007 report the IPCC estimated that by the end of this century the atmosphere could contain the equivalent of 600 to 1,500 ppm of carbon dioxide. These figures include other greenhouse gases by converting the warming produced by them into the number of carbon dioxide molecules needed to produce the same warming.

The report focuses on greenhouse gases because they are the main human contribution to climate change, but it also includes effects of other changes, such as in land use, on the atmosphere.

Let's turn now to one of the most prominent examples of Earth's changing climate: the warming of the Arctic.

The significance of polar ice

The chapter-opening story based on Charles Wohlforth's book, *The Whale and the Supercomputer*, shows how the general loss of Arctic Ocean sea ice affects the Iñupiat who live on the Arctic Ocean coast. In his book, Wohlforth weaves together the stories of the Iñupiat and some of the many scientists who are trying to understand the links between the Arctic's weather and ice and a changing climate.

The Iñupiat and other native peoples around the Arctic are worried because warming is making their traditional knowledge of the ice and weather a less reliable guide to staying alive in a harsh world. In the opening of Chapter 2, we saw that scientists are concerned about melting polar ice because, among other things, this will allow the earth to absorb heat that ice now reflects away. When ocean water that was once covered by sea ice absorbs

solar energy the ice used to reflect away, the ocean warms and in turn warms the air above it. This is called a positive feedback because it feeds on itself; the more the ice melts, the more solar heat the ocean absorbs, and the more ice the heat can melt.

This is one example of why the Arctic, which is far away from the parts of the earth where most people live and where most farms, ranches, orchards, and beehives are located, is important to life on the planet. We can think of the polar regions as being like Earth's air conditioners. The Arctic and Antarctic do the same things an air conditioner in a building does: they lower the temperature and reduce the water vapor in the air. Large amounts of the water vapor end up in the glaciers and ice sheets that contain most of the earth's freshwater, as we saw in Chapter 4. In addition to cooling the air, the Arctic and Antarctic also cool the warm water that ocean currents transport poleward, sending the coldest water back toward the equator as Arctic and Antarctic deep water.

If a building's air conditioner isn't fully functioning, the building warms. When Earth's air conditioners become less efficient, the entire earth is likely to grow warmer. This is, in a general way, one reason why climate scientists are concerned with what happens in the polar regions. In more specific ways, climate scientists want to learn how polar changes can affect the strength and paths of middle-latitude storms. Since these storms draw their energy from temperature contrasts, changes in polar temperatures could affect them. What the effects are likely to be isn't clear. For instance, one study using a climate model that includes ocean-atmosphere interactions showed that a particular warming scenario would suppress the east Asian monsoon and bring more precipitation to the U.S. West Coast. But, notes Mark Serreze, a senior research scientist at the National Snow and Ice Data Center in Boulder, Colorado, another suggests that losing the ice could lead to drought in the U.S. West.

Warming and the Arctic. While the reduction in Arctic Ocean sea ice is one of the most striking changes in the far north, big changes are also occurring on land. Residents and scientists have noted that snow isn't on the ground quite as long each year, winters and summers tend to be warmer, wildfires are becoming more of a problem in

Antarctic sea ice

The annual winter growth and summer melting of the sea ice around Antarctica are the world's largest oceanic-atmospheric events caused by seasonal changes. The annual ice growth and melting both affect global climate.

The white area is covered by sea ice. The violet lines show the average 1979–2000 ice coverage.

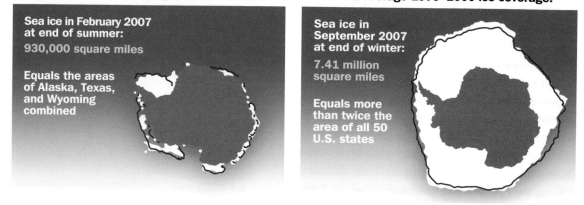

Sea ice in February 2007 at end of summer:
930,000 square miles

Equals the areas of Alaska, Texas, and Wyoming combined

Sea ice in September 2007 at end of winter:
7.41 million square miles

Equals more than twice the area of all 50 U.S. states

the days lengthen and temperatures rise, it begins to melt slowly. The much larger amounts of sea ice in the ocean around Antarctica also come and go with the seasons. In fact, the formation and melting of sea ice in both the Arctic and Antarctic are the largest seasonal climatic events on Earth.

Different in the south. While the average amount of summer Arctic sea ice has been decreasing by approximately 3 percent per decade since 1979, National Snow and Ice Data Center scientists say ice around Antarctica has been increasing by less than one percent a decade. While the measurements aren't precise enough to say that Antarctic sea ice is growing, it's not very likely to be shrinking in the near term like Arctic sea ice.

Snow and Ice Data Center scientists give three reasons why Antarctic sea ice isn't shrinking like ice at the other end of the earth. First, to a large extent, the ocean currents and winds that circle Antarctica isolate the continent and the ocean around it from being warmed as much as the Arctic is by warm air and ocean currents. Second, the cold, katabatic winds that blow from the Antarctic continent over the ocean help keep the ocean surface cool. Finally, all but approximately 20 percent of the sea ice that forms around Antarctica in the winter melts by the end of summer, while in the Arctic approximately 40 percent of the sea ice normally lasts through the summer.

While the Arctic has been losing some winter sea ice, the big decline is in the summer. Since most of the sea ice around Antarctica melts in the

summer anyway, "you wouldn't expect to see as much of a trend," says Walt Meyer of the Snow and Ice Data Center. This doesn't mean a warming world isn't affecting Antarctica; it's just not affecting the earth's southernmost end as soon as other regions. In fact, Ted Scambos of the Snow and Ice Data Center calls Antarctica "the sleeping giant," which, with continued climate prodding, could soon stir.

When the Rhode Island–size Larsen B Ice Shelf on the Antarctic Peninsula disintegrated between January 31 and March 7, 2002, some news stories implied that it was a sign that global warming was melting Antarctica's ice. But, unlike large parts of the Arctic, most of the Antarctic continent is not yet warming up, as far as can be determined from the spotty temperature records. In fact, some parts seem to have cooled a little. But the peninsula—the part of Antarctica that looks like an arm reaching for the tip of South America—is warming significantly.

Floods from melting ice

The possibility that global warming will melt the huge amounts of ice covering Antarctica and Greenland is one of the major climate change concerns. During the last half of the twentieth century, glaciers in many parts of the world began melting. At the current rate, all of the glaciers in Glacier National Park in Montana could melt by the middle of this century. While the loss of such glaciers often poses huge problems for the areas where they are located, water from melting glaciers contributes relatively little to global sea levels.

From 1993 to 2003, global sea levels rose by approximately 3.1 millimeters (0.12 inches) per year from the melting of mountain glaciers, a little melting from the Antarctica and Greenland ice sheets, plus the expansion of seawater as it warms (like most substances, water expands as it warms).

In its 2007 report, the IPCC says that warming during this century is expected to speed up melting of ice from glaciers and from Greenland and Antarctica, and this could raise sea levels by amounts ranging from approximately half a foot to nearly two feet by the end of this century. Some climate scientists think the IPCC underestimated the potential for sea-level rise because not enough is known about how the Antarctic and Greenland ice sheets work.

While two feet doesn't seem like much of a sea level rise, it would be enough to increase the storm surge damage from tropical cyclones and other storms. In fact, many ordinary storms that aren't a concern now would cause coastal flooding. The biggest concern is really over a much longer term; glaciers and ice sheets could begin melting faster in the future as the climate continues warming.

As sea levels rise, even a few inches, salt water pushes farther up rivers. The U.S. Environmental Protection Agency says such salt-water intrusion would affect New York City, Philadelphia, and much of California's Central Valley because much of their freshwater comes from sections of rivers that are only a little upstream from where their water becomes salty during droughts. Rising sea levels would also push more salt water into the aquifers that supply freshwater to coastal areas.

The ice-oceans balance. In the 2007 IPCC report, estimates of sea level rise are based on how much carbon dioxide (CO_2) could be added to the atmosphere this century and the "contribution [of water to the oceans] due to increased ice flow from Greenland and Antarctica at the rates observed for 1993–2003," but future rates could be different than those observed during this period.

How much of the earth's water is locked up as ice depends on the balance between evaporation of water from the oceans that ends up falling on glaciers and ice sheets as snow and how fast the glaciers and ice sheets melt. When falling snow more than makes up for melting, glaciers and ice sheets grow. When falling snow does not make up for melting, glaciers and ice sheets add water to the world's oceans and sea levels rise.

As we saw in Chapter 4, the warmer the air, the more water vapor is needed to saturate it. This means that as the air warms up, more snow can fall onto glaciers and ice sheets. For example, if the air's temperature increases from 0 to 15 degrees Fahrenheit, the amount of water vapor needed to saturate the air—that is, make the relative humidity 100 percent—almost doubles, from 0.93 to 1.84 grams of water vapor per kilogram of air. Snow falling from the warmer air onto the highest parts of Greenland's ice sheet or almost anywhere on the Antarctic Ice Sheet would not melt because the air's temperature stays below 32 degrees Fahrenheit for almost all of the year on the highest parts of Greenland's ice sheet and always over almost all of Antarctica.

The IPCC report expects Greenland to continue contributing to sea level rise during this century. Meanwhile, "Current global model studies project that the Antarctic Ice Sheet will remain too cold for widespread surface melting and is expected to gain in mass due to increased snowfall." But, the IPCC warns, Antarctica could begin losing ice "if dynamical ice discharge [described below] dominates the ice sheet mass balance."

Dynamics of ice sheets. "Dynamical ice discharge" refers to the fact that ice sheets don't flow like a single glacier but have "streams" of ice that move faster than the parts of the ice shelf on either side.

One of the biggest challenges facing glaciologists is to better understand all of the factors that control how fast ice sheets move. What's at the bottom of the glacier or ice sheet makes a difference. If the ice is frozen to the rock at its bottom, it will move much slower than ice that's sitting on a thin layer of water. This and the **till**—clay, sand, and rocks of different sizes and shapes under the ice— enable the ice to move faster.

Ice shelves could also play a role in the dynamics of ice sheet balance. Ice shelves form when a glacier or ice sheet pushes out over water without the ice breaking off. The resulting ice shelf is floating even though it's still attached to the ice sheet. Glaciologists believe that some ice shelves are holding back ice on land. When the ice shelf weakens or

The melting sea ice in the Arctic Ocean doesn't cause sea levels to rise because the ice is already floating. If you want to see how this works, fill a glass to the brim with ice and water and let the ice melt—the glass will not overflow.

of the upper atmosphere. Still, scientific understanding was firm enough for the nations of the world to agree to reduce and eventually end production of the substances most responsible for ozone depletion. The fact that substitutes for CFCs were developed is very important—imagine what would be happening to the ozone layer today if the only choice had been between stopping ozone destruction or giving up refrigeration and air conditioning.

Climate change caused by global warming is a vastly more complex problem. Despite what you sometimes hear, by the first years of the twenty-first century, the overwhelming majority of scientists who specialize in climate agree on the main point: The climate is changing and humans are, at least to some degree, responsible. They agree because a growing amount of evidence of different kinds convinced even those who were skeptical—as scientists should be—that humans are helping to cause the earth's average temperature to warm. The disagreements among climate scientists are about the details. These are sometimes important details, such as the relative importance of the effect of greenhouse gases and land-use changes on climate.

The joint statement that the national academies of science of eleven nations, including the United States, adopted in June 2005 sums up the science: "The scientific understanding of climate change is now sufficiently clear to justify nations taking prompt action…A lack of full scientific certainty about some aspects of climate change is not a reason for delaying an immediate response that will, at reasonable cost, prevent dangerous anthropogenic interference with the climate system."

The core of the actions needed to respond to human-caused climate change is to reduce the amount of carbon dioxide going into the air from the burning of fossil fuels. This is vastly more difficult than cutting back on releases of the substances that harm ozone, because fossil fuels power much of the world economy. The attempt to fashion a climate-change equivalent to the Montreal Protocol on harmful ozone-destroying substances illustrates the difficulty of addressing climate change, as seen in the next section.

The Kyoto treaty. The IPCC's 1995 conclusion that "the balance of evidence suggests that there is a discernible human influence on global climate" led government leaders from many nations to try to reach an agreement that would do for climate change what the Montreal Protocol did for ozone destruction. More than 6,000 official delegates and hundreds of additional people, including representatives of environmental groups and reporters, converged on Kyoto, Japan, in December 1997 to debate and try to forge a treaty limiting greenhouse gas emissions.

Representatives of the United States proposed that industrial nations aim to gradually reduce their greenhouse gas emissions to 1990 levels, while Western European nations wanted to reduce the emissions even more. China and many developing nations argued that they should be exempt from any limits because they needed to catch up with the industrialized nations in terms of building an industrial base.

As the conference dragged on, the odds grew that everyone would go home without a treaty until Al Gore, who was then U.S. Vice President, flew in on the last day and managed to talk the delegates into agreeing on the Kyoto Protocol, which would have wealthy nations cut greenhouse gas emissions by 2010.

The U.S. Senate has to ratify all treaties and the day after agreement was reached in Kyoto, Republican congressional leaders said at a press copference that the protocol would be "dead on arrival" at the Senate. The Clinton administration then said it would not send the treaty to the Senate for ratification unless the largest developing countries agreed to reduce greenhouse emissions. This didn't occur, and in March 2001 the new president, George W. Bush, rejected the treaty.

To go into effect, the Kyoto Protocol had to be ratified by developed nations whose collective share of global greenhouse gas emissions added up to 55 percent of the global total. This finally happened on November 5, 2004, when Russian President Vladimir Putin signed the treaty. Russia's legislature approved the treaty and Putin signed it after the European Union agreed to endorse Russian membership in the World Trade Organization. Also, Russia expected no problems reducing its greenhouse emissions below 1990 levels as the treaty required, since the notoriously polluting factories and power plants of the former Soviet Union that had been pumping out greenhouse gases for decades fell victim to the collapse that took place when the Soviet Union fell apart in 1991.

Making sense of weather, climate

In this book, we haven't attempted to discuss what might be done about climate change. Many people are making suggestions that range from buying more energy-efficient light bulbs to developing whole new ways to power the global economy, beginning with renewable energy sources such as **biofuels** and capturing and storing carbon deep underground. There are even speculative proposals to block some solar energy from reaching the earth.

Like the details of climate change, these are topics of entire books. Many more such books will be published in the coming years. Global warming is a topic of political debate, with its own vast literature both in print and on the Web. Our aim in this book, not only in this chapter but throughout, is to help you understand the basic science underlying weather and climate. With such an understanding you'll be able to make more sense of both the daily weather forecast and the debate over climate change.

As citizens of the world, we should be more than bystanders as policy makers decide whether to do anything significant about the threat of climate change. As the American Meteorological Society statement on climate change adopted in February 2007 indicates, these decisions will be difficult: "Policy decisions are seldom made in a context of absolute certainty. Some continued climate change is inevitable, and the policy debate should also consider the best ways to adapt to climate change. Prudence dictates extreme care in managing our relationship with the only planet known to be capable of sustaining human life."

An understanding of weather and climate can both help individuals to make sensible personal decisions and urge elected officials to make better public decisions about preparing for weather events, such as hurricanes, floods, tornadoes, heat waves, and more. All of these will continue to occur no matter how the climate changes, although some such as heat waves could become more frequent. Below we look at two of the many weather-related dangers facing us in the coming years: the potential

for increasingly severe storm surge and wind damage associated with hurricanes and floods in densely populated areas lacking strong flood protection.

Hurricane advice from scientists. In July 2006 ten leading hurricane scientists on both sides of the previous year's debate about warming and hurricanes issued a joint statement saying the debate over warming and hurricanes is not as important as "the ever-growing concentration of population and wealth in vulnerable coastal regions."

The scientists said that as more people move into coastal areas, hurricanes will become more deadly and expensive whether or not climate change affects storms. Scores of scientists and engineers had warned of the threat to New Orleans long before climate change was seriously considered, and a Katrina-like storm or worse was inevitable and could happen again even in a stable climate.

The scientists criticized government policies that "serve to subsidize risk." They said that further research will eventually answer questions about the relation between warming and tropical cyclones, but the big problem that needs to be addressed now is the growth of coastal populations in locations where hurricanes are most devastating.

After the damages to the Gulf Coast from hurricanes Katrina and Rita in 2005 and to Florida from Wilma that same year and from hurricanes Charley, Frances, Ivan, and Jeanne in 2004, many people in those areas will probably go a few years before they again become blasé about hurricanes. If coastal development along the U.S. East and Gulf coasts slows, it may be because of rising insurance rates even more than because individuals become reluctant to move into the possible path of the next Katrina.

With or without global warming, hurricanes are a potential danger to all of the U.S. East Coast, including New York City. In fact, the city's Office of Emergency Management says, "New York's densely populated and highly developed coastline makes the city among the most vulnerable to hurricane-related damage." They also say a major hurricane could push more than 30 feet of storm surge into some parts of the city. While the odds are smaller

A science advisor to Congress

U.S. senators and representatives vote on matters ranging from the armed forces to zoology, says Ana Unruh Cohen, a scientist who's turned to a public policy career. They rely on the expert knowledge of their staff, such as Unruh Cohen, deputy staff director for the House Select Committee on Energy Independence and Global Warming.

For instance, for a May 2007 hearing on the "Economic Impacts of Global Warming: Green Col-

Ana Unruh Cohen likes "being a bridge from the scientific community to policy makers and the public."

lar Jobs," Unruh Cohen says she "did it all…conceived the idea, recruited the witnesses, wrote the chairman's opening statement, and (with the help of my colleagues) put together questions" to ask the witnesses.

During the hearing, she and other staff members sat behind committee members sitting at the long table in the front of the hearing room. A few times during the hearing she quietly conferred with the chairman, Edward J. Markey. "We are back there before every hearing," she says. "You never know what will come up. Sometimes if a witness brings up an issue that catches the chairman's attention, he'll ask us to do more research," she says. The chairman might "ask about how to respond to an issue the Republicans have brought up or what questions he should ask the next witness."

Unruh Cohen thinks that in the future, "political science graduate students will be writing theses on how we got our first climate-change law.

This will be a critical time for them to examine. It's interesting to be in the thick of it."

While growing up in Corpus Christi, Texas, Unruh Cohen spent time watching birds with her parents and camping, which led to a love of the outdoors and concerns about the environment. She was also "really good at chemistry. My Dad was a chemist, and I think I just picked it up…he talked about his work around the dinner table."

While earning her bachelor's degree in chemistry at Trinity University in San Antonio, Texas, she realized that a lot of chemistry is involved in solving environmental problems. After graduating with a bachelor's degree in chemistry from Trinity, she earned her doctorate in geochemistry from Oxford University in England as a Rhodes Scholar.

She came to Washington in 2001 as an American Meteorological Society Congressional Science Fellow under a program that puts scientists to work for a member of Congress for a year. Unruh Cohen worked for Markey, a Massachusetts Democrat, as a fellow and then as a staff member until 2004.

The fellowships are "kind of a safe way to try out if you like policy; it's not totally severing your academic ties. There was a chance to go back to academia if I was not happy." But, "I just loved it."

After two and a half years as director of environmental policy at the Center for American Progress in Washington she returned to Capitol Hill in 2007 after the Democrats regained control of Congress, created the House Select Committee, and named Markey chairman.

"One of the best lessons and also probably one the hardest lessons I've learned up on Capitol Hill is that in the end, politics does trump substance," she says. "The critical thing for scientists and the science community to understand is that the big challenge for us is to make the substance politically relevant."

While she thinks a few of the politicians she works with "are in it for money and power, my gut feeling is that for the most part, even representatives I didn't agree with politically are doing it because they want to serve the country. We both look at the same America and we come to different conclusions about what policies should be put into place to solve problems we might both agree on."

that a hurricane will hit New York City than Miami or Houston, or New Orleans again, the city's huge population increases the consequences.

The danger of floods. A warming climate could bring more intense rain to some areas, making floods worse, while other areas become drier. We can be sure, however, that no matter what effects a warmer climate has on rainfall, the earth will still experience floods. We can also be sure that human decisions will make these floods more destructive and deadly in places. Katrina's floods in New Orleans are a perfect example of how things that people do, or don't do, can make weather events worse. Poorly designed, constructed, and maintained levees were as responsible as Hurricane Katrina for the city's floods.

Sacramento, California, is another major city that faces a danger of serious flooding. In fact, says the Sacramento Area Flood Control Agency, "Sacramento's risk of flooding is the greatest of any major city in the country." The August 13, 2007, issue of *Business Week* magazine described the levees that protect large areas in and around Sacramento: "Many...are simply mounds of river muck, piled originally by shovel and wheelbarrow in the 1800s and never designed for long-term protection of homes and other structures. Moreover, the drained land has subsided, leaving huge tracts 15 or 20 feet below sea level."

In addition, the Flood Control Agency says, the amount of water that flows out of the Sierra Nevada range during floods appears to be increasing. When engineers designed the Folsom Dam on the American River in 1950, they used records of the river's flow and statistical analysis to predict the size and frequency of large floods. Their goal was to have the dam hold back enough water to ensure floods wouldn't endanger the levees downstream, including those in Sacramento. The dam was built to be able to store the extra water from a flood with one chance in 500 of occurring in any year.

A record flood occurred in 1951 while the dam was under construction. It was larger than any flood in the previous forty-five years. In 1956, with the dam nearly completed, another record flood occurred. The dam's engineers had predicted the river would need a year to fill the lake behind the dam, but the 1956 flood filled it in a week, saving Sacramento from flooding. Other record floods occurred in 1964, 1986, and 1997. What the Folsom Dam's designers thought to be a flood with a one in 500 chance of occurring during any year had turned out to be a flood with a one-in-fifty chance of occurring in a given year.

In January 2007, the Associated Press reported that developers in a flood-prone Sacramento area were giving prospective homebuyers a DVD about the possibility of flooding in the area. The developers said they'd rather lose a buyer or two than have people move in without knowing of the flood danger. This is an example of how people other than public officials can take responsibility for being prepared for what the weather might bring. Public officials have also acted. On February 24, 2006, California Governor Schwarzenegger declared a state of emergency for the levees that protect the Sacramento area, saying that a Katrina-like disaster was possible. In November 2006, the state's voters approved a $4.1 billion bond issue to pay for strengthening levees and improving flood control.

The personal side of weather

In the opening of Chapter 1, we saw how Robert Ricks, a National Weather Service meteorologist, and his extended family fared after Ricks issued a warning that he hoped would spur people to flee New Orleans as Hurricane Katrina approached.

"My family was away from the immediate impact area and was able to regain normalcy rather quickly," he said in December 2006, sixteen months after Katrina demolished New Orleans and the Mississippi coast.

While Katrina disrupted their lives, Robert Ricks and members of his extended family were fortunate compared to thousands of other victims of the storm.

"Yesterday I went down into the city. You still see the water line on the houses," Ricks said in December 2006. "You can see that large portions of the city are still uninhabitable. Sixteen months removed, and not a day goes by without something in the news about Katrina. You go around and you see FEMA trailers everywhere; it's a lifestyle people are just getting tired of. In some areas you see improvements, but for every area of hope there are two areas of despair."

In Chapter 8, we saw how decisions by Bob and Terri Parsons, owners of the Parsons Company in Roanoke, Illinois, ensured the survival of not only everyone in the building but of the business—and its jobs—when a tornado flattened the factory in 2004. Individuals, businesses, and other institutions such as schools can make similar decisions that will save lives and help the business or institution survive devastating but relatively small-scale disasters such as tornadoes, flash floods, and wildfires.

Individuals can also take steps to ensure their own and their family's survival when a hurricane moves ashore or any other kind of weather disaster threatens their homes. But only government can take the steps needed to avoid the kind of aftermath that Ricks described in New Orleans at the end of 2006. "The economic engine is still out of kilter. There's not enough labor, not enough service employees to support businesses. They can't get enough new people to move in because there's not enough housing. The economic engine is not an engine, it's more like a two-cylinder lawn mower."

You can be sure that over the coming years you are going to be hearing more about the need to address human actions, especially burning fossil fuels, that are affecting the climate. Even if we had no concerns about the effects of a changing climate, governments and individuals would still face decisions about the best ways to cope with future hurricanes, floods, droughts, wildfires, heat waves, and air pollution. In each case, individuals and governments will have to make decisions about the proper balance between investments, such as for improved flood protection, and regulating what are usually considered to be private matters, such as where people are allowed to build homes. We hope that what we've communicated in this book will help you become a more informed participant in private and public discussions and decisions involving weather and climate. We also hope it will help you better appreciate both the science and beauty of the natural world, beginning with the sky and the weather in all of its moods.

GLOSSARY

A

acid rain: Rain that is more acid than natural rainwater, usually due to the presence of sulfuric acid or nitric acid, often from human sources.

active layer: The thin layer of soil atop **permafrost** in polar regions or at high altitudes, which is frozen during winter and thaws only during a brief summer season, allowing small plants to grow.

active remote sensing: Any technology for acquiring data about an object or phenomenon using a device that emits energy, which is then scattered or reflected by the object or phenomenon of interest. The device receives the scattered energy and converts it into data about the phenomenon. Opposite of **passive remote sensing**.

adiabatic: Heating or cooling of a gas without any heat being exchanged with the surroundings. In meteorology, generally refers to the cooling of rising air and the heating of sinking air.

adiabatic lapse rate: The change in air temperature as the result of adiabatic heating or cooling as the air moves up or down.

adiabatic winds: A term that is sometimes used for **foehn winds**, which warm the region that they blow into. The opposite of a **bora wind**. However, all winds warm adiabatically as they flow downhill, including bora winds.

advection: Horizontal transport of atmospheric properties, such as heat or humidity, by the wind.

aerosol: A liquid or solid particle suspended in a gas such as the air.

air parcel: An imaginary volume of air to which atmospheric scientists may assign any of the basic properties of atmospheric air. Often used as a convenient tracer of air movements or atmospheric processes. Similar to **water parcel**.

air pollution: The presence of particles or gases in the air that could harm plants or animals.

Antarctic Bottom Water: Cold, salty water that sinks to the ocean floor off the Antarctic coast and flows along the ocean bottom into the Northern Hemisphere.

Antarctic Oscillation: The dominant pattern of atmospheric pressure and wind changes in the Southern Hemisphere that is not associated with seasonal changes. Also known as the **Southern Annular Mode**.

anthropogenic: Caused by or produced by humans.

apparent temperature: A measure of the effects of high air temperatures and high relative humidity on human comfort and safety.

Arctic Oscillation: An atmospheric circulation pattern in which the atmospheric pressure over the Arctic varies in opposition with that over middle latitudes on time scales of weeks to decades. It affects the strength of the westerly winds that encircle the Arctic. Same as **Northern Annular Mode**

atmosphere: The envelope of air surrounding a planet and bound to it by the planet's gravitational attraction.

atmospheric river: A narrow stream of very humid air moving from the tropics into the middle latitudes of either hemisphere.

B

biofuels: A solid, liquid, or gas fuel derived from plants or animals.

biosphere: The parts of the earth's land, oceans, and atmosphere inhabited by living plants or animals. Sometimes used to refer to all of the planet's living things.

blizzard: Severe winter weather characterized by strong winds carrying blowing snow. The NWS specifies that a blizzard must have winds of 35 mph or higher, low temperatures (no set value), enough blowing snow to reduce visibility to less than 0.25 miles, and last for an extended period of time (at least three hours). In popular use in the United States and the United Kingdom, often used for a heavy snowstorm with strong winds.

bora (wind): A gravity-driven downslope wind with a source so cold that the air cools the low lands in spite of compressional warming.

bow echo: A bow-shaped line of thunderstorms that shows up most clearly on weather radar and which is often associated with swaths of damaging straight-line winds and small tornadoes.

C

cap and trade (pollution control): A pollution-abatement system under which a government or another authority sets a limit, called a "cap," on the amount of a particular pollutant that can be emitted. Polluters, such as manufacturing plants are given permits, called allowances or credits, for the right to emit certain amounts of pollution. A company that is emitting less pollution than its limits can sell the credits for the pollutants it is not emitting to a company that can't (or doesn't choose to) meet its limit. "Trade" refers to the selling and buying of credits.

capacitor (electrical): A device that stores electrical charge on conductive material separated by an insulator.

CAPE: See **Convective Available Potential Energy**.

capping inversion (cap): While "inversion" implies that temperature increases with height, the term "capping inversion" is used more loosely for any stable layer in the lower atmosphere that inhibits development of convection.

carbon dioxide (CO_2): A colorless gas at temperatures and pressures in the atmosphere. Each molecule consists of one carbon atom and two oxygen atoms. It is a **greenhouse gas**.

Chinook winds: A **foehn wind** that blows down the eastern slopes of the Rocky Mountains in North America.

civil twilight: The interval of incomplete darkness between sunrise or sunset and the time when the center of the sun's disc is six degrees below the horizon. Generally the minimum sky illumination required to carry on normal outside activities without artificial light.

cloud microphysics: Cloud processes, such as growth and evaporation, that take place on the scale of the individual aerosol or precipitation particle as opposed to the scale of the visual cloud.

Definitions of terms in this glossary, with a few exceptions, are based on the *Glossary of Weather and Climate with Related Oceanic and Hydrologic Terms*, Ira W. Geer editor, published by the AMS, Boston, 1996; *Online Ocean Studies*, AMS Education Program, published by the AMS, Boston, 2005; and the online *Glossary of Meteorology*, second edition (which is more technical), published by the AMS. Definitions of terms connected with Earth's ice are generally based on definitions in the National Snow and Ice Data Center's online *Cryosphere Glossary*.

Words in **boldface** refer to other definitions in this glossary.

cloud seeding: Any technique that adds certain particles to a cloud with the intent of altering the cloud's natural development in ways that will lead to precipitation or increased precipitation.

condensation nuclei: Liquid or solid particles in the air upon which water vapor begins to condense.

conductor (electrical): A substance that will carry an electrical current.

convection: In meteorology and oceanography, the up and down motion of air or water caused by temperature differences.

convective: Resulting from **convection**, such as convective clouds. Also referring to clouds and precipitation depending on convection, especially **thunderstorms**.

Convective Available Potential Energy (CAPE): The maximum energy that could be released by air parcels rising to the height at which their temperature equals the environmental temperature.

convergence: In meteorology and oceanography, a pattern of horizontal flow of air or water that brings about a net inflow of fluid into a region. The opposite of **divergence**. Convergence is accompanied by a compensating vertical motion.

Coriolis force: The force in the frame of reference of a rotating planet caused by the planet's rotation, which causes the deflection of moving objects in relation to the planet's surface.

corona: In the atmosphere, one or more colored rings with a small radius around the disk of the sun, moon, or other light that is veiled by a thin cloud and caused by **diffraction**.

covalent bonds: A form of chemical bonding characterized by the sharing of one or more pairs of electrons between atoms that make up a molecule.

crosswind: A wind, or component of a wind, directed perpendicularly to the direction of travel of an exposed, moving object, such as an aircraft.

cryrosphere: The part of the Earth covered by permanent ice.

cumulus cloud: A type of cloud characterized by individual, vertically developed elements.

cut-off low: An area of low atmospheric pressure that has become separated from the westerly flow of air.

cyclone: A weather system characterized by a relatively low-pressure center and winds that circle around the low pressure in a counterclockwise manner in the Northern Hemisphere, and clockwise in the Southern Hemisphere.

D

data assimilation: The combining of diverse data, possibly sampled at different times and intervals and different locations, and from different types of sources, such as surface instruments and radar, into a unified and consistent description of a physical system, such as the state of the atmosphere.

deposition: In meteorology, the change of water vapor directly into ice without first becoming a liquid.

deposition nuclei: Small particles that hasten the deposition of water vapor into ice.

derecho: (Pronounced "deh-RAY-cho" in English) A widespread and long-lived windstorm associated with a band of rapidly moving showers or thunderstorms.

dew point: The temperature to which a parcel of air must be cooled at constant pressure and constant water vapor content in order for **saturation** to occur.

diffraction: The bending of any kind of waves around objects. In the atmosphere, the bending of visible light waves can create light, dark, or colored bands such as cloud **coronas** and **iridescence**.

divergence: In meteorology and oceanography, a pattern of horizontal flow of air or water that produces a net outflow of fluid from a region. The opposite of **convergence**. Divergence is accompanied by a compensating vertical motion.

doldrums: A nautical term for the equatorial trough that refers to the normally light and variable winds within five to ten degrees of latitude from the equator.

Doppler radar: In meteorology, radar that detects wind speeds and directions in addition to **hydrometeors** by determining the speed of reflecting particles toward and away from the radar sing the **Doppler effect**.

downburst: An exceptionally energetic downdraft from a shower or thunderstorm that spreads out across the Earth's surface as strong and gusty horizontal winds.

downdraft: A downward-flowing air current in clouds or clear air.

dropsonde: A package of instruments designed to drop from an airplane by parachute to measure and radio back data on temperature, humidity, atmospheric pressure, and wind speed and direction.

dry adiabatic lapse rate: The rate at which rising air cools if none of its water vapor is condensing, or sinking air warms; 5.4 degrees F per 1,000 feet.

dryline: A boundary between masses of humid air and dry air, which is often a site for development of thunderstorms.

dry thunderstorms: Thunderstorms that commonly form in arid regions with bases high above the ground with rain that evaporates before reaching the ground. Lightning from such storms is a common cause of wildfires in the western United States.

dynamical forecasts: Weather predictions based on solving the mathematical equations that govern the atmosphere.

E

echo top: The highest precipitation detected by weather radar in a cloud. The actual visual cloud top is normally higher but made of ice crystals or water droplets too small for radar detection.

electromagnetic radiation: A form of energy propagated through space or material media by advancing **transverse wave** disturbances in electric and magnetic fields.

electromagnetic spectrum: The ordered array of all known types of electromagnetic radiation, arranged by wavelength.

electrometeors: Any visible or audible indicator of atmospheric electricity including lightning, thunder, and auroras.

El Niño: The chain of oceanic-atmospheric events characterized by anomalous warming of surface water of the eastern, tropical Pacific and cooling of the surface water in the western Pacific. These changes affect patterns of atmospheric pressure at the surface and aloft with resulting changes in winds that affect weather as far away as Africa.

El Niño neutral state, also **El Niño neutral:** The phase of the ENSO cycle characterized by tropical Pacific water temperatures between the extremes typical of El Niño and La Niña.

ensemble forecast: A forecast produced by making multiple runs of a numerical forecasting model with slight changes in the initial conditions to see how much the results differ.

ENSO: Acronym for El Niño-Southern Oscillation, coined in the early 1980s in recognition of the intimate linkage between **El Niño** events and the **Southern Oscillation**. The global ocean-atmosphere phenomenon to which this term applies is sometimes referred to as the "ENSO cycle." ENSO is often used to refer to the entire cycle that includes the El Niño, **La Niña**, and neutral phases.

entrainment: In meteorology, the mixing of environmental air into a pre-existing, organized air current so that the environmental air becomes part of the current.

environmental lapse rate: The rate of decrease or increase of the temperature of air that is not moving up or down. Contrast with the **adiabatic lapse rate**.

evapotranspiration: The vaporization of water through direct evaporation from wet surfaces plus the release of water vapor by plants through leaf pores (transpiration).

extratropical cyclone: Any **cyclone** larger than the **mesoscale** that is not a **tropical cyclone**; characterized by different air masses separated by fronts. Most often used for middle-latitude cyclones.

extremophile: An organism that lives in physically or geochemically extreme conditions, such as extremely low or high temperatures, high salinity, or acidity, that are deadly to most life on Earth.

eye: In meteorology, the center of a storm, usually a tropical cyclone, that is at least partly cloud free.

eyewall: The ring of thunderstorms that surrounds or partly surrounds the eye of a tropical cyclone. Usually the location of the storm's strongest winds.

eyewall replacement cycle: A cycle, which takes several hours, characterized by formation of an outer eyewall around the eyewall of a tropical cyclone and the weakening and dispersion of the original eyewall with a weakening of the storm's strongest winds. The new eyewall then can shrink in diameter as the storm regains strength.

F

Ferrel cell: The average atmospheric circulation in the middle and high latitudes, poleward from the Hadley cell, of both hemispheres.

First proposed by William Ferrel (1817–1891), an American meteorologist.

firn: Snow pack consisting of rounded, well-bonded snow crystals older than one year.

fjord: A long, narrow inlet from the ocean with steep hills or mountains on both sides.

flash (lightning): The total observed luminous phenomenon caused by a lightning discharge. Usually made of several **strokes**.

flood: The overflowing of the normal confines of a stream or other body of water, or the accumulation of water over areas that are not normally submerged.

foehn winds: Often used to describe any wind that warms and dries as it flows downhill. Originally, the warm, dry winds in alpine valleys of Austria and Germany.

forensic meteorology: The collection and analysis of meteorological data for possible use in legal matters, such as expert testimony at a trial.

fossil fuel: Hydrocarbons such as coal, oil, and natural gas formed from the remains of plants and animals buried deep in the earth millions of years ago.

freezing nuclei: Particles that when present within a mass of **supercooled water** will initiate the growth of an ice crystal around themselves.

frost: The fuzzy layer of ice crystals on a cold object, such as a window or bridge, that forms by direct deposition of water vapor to solid ice. Also, the condition that exists when the temperature of the earth's surface and earthbound objects fall below freezing.

frostbite: The freezing of parts of the body of a human or animal, causing damage to tissue.

G

geoscientists: Scientists who study aspects of the planets, especially the earth, including but not limited to the **atmosphere**, the **cryosphere**, the **hydrosphere**, and solid matter such as soil and rocks, or the chemistry, physics, and history of any of these.

glacier: A mass of ice, usually larger than one tenth of a square kilometer, that originates on land and is made of compacted snow. It must be flowing slowly in response to gravity or have flowed in the past from the area where the snow accumulated to where it melted on land or flowed to water where it melted or pieces broke off as icebergs.

glacier ice: Well-bonded ice crystals compacted from snow with a density greater than 55 pounds per cubic foot.

Global Forecast System model: One of the several **numerical weather prediction models** run by the U.S. National Centers for Environmental Prediction and by government and private forecasting services in several nations.

GPS: The Global Positioning System of 24 satellites that supply precise navigation information, which is used not only by operators of aircraft, ships, and motor vehicles but also for research into movements of ice sheets and ocean currents.

greenhouse effect: The warming of the earth by **greenhouse gases** in the atmosphere.

greenhouse gases: Gases that absorb and re-radiate radiation from the earth.

grid: In meteorology, the collection of points, usually uniformly spaced, to which the variables used in a **numerical weather prediction model** apply.

ground truth: Data obtained directly about a phenomenon being observed by remote-sensing technology, which allow scientists to calibrate the remote-sensing technology by comparing the two kinds of observations.

H

Hadley cell: The tropical air circulation consisting of surface winds moving toward the equator in both hemispheres, rising in convective clouds, and flowing aloft toward the poles. Named for George Hadley, a British scientist who described it in 1735.

hail (hailstone): Ice that falls from thunderstorms with strong updrafts as balls or irregular lumps. Contrast with **sleet**.

halo: Any **photometeor**, including rings, arcs, pillars, or bright spots, around the sun or moon caused by clouds made of ice crystals or by ice crystals floating in the air.

halocline: A layer of ocean water in which the salinity increases rapidly with depth.

Heat Index: The **apparent temperature** that describes the combined effect of high air temperatures and high humidity levels, which reduce the body's ability to cool itself.

hot tower: A tropical cumulonimbus cloud that penetrates into the lower stratosphere.

hurricane: A **tropical cyclone** with sustained surface winds in excess of 74 mph in the North Atlantic Ocean, Caribbean Sea, Gulf of Mexico, and in the eastern and central North Pacific east of the International Date Line. **Tropical cyclones** are called **typhoons** or **cyclones** in other parts of the world.

hydrocarbon: A compound consisting of hydrogen and carbon atoms.

hydrogen bond: The attracting force between molecules of a substance with opposite electrical charges on different sides and which includes hydrogen atoms.

hydrometeorological: Referring to aspects of **meteorology** concerned with water such as flooding, hydrological power, irrigation, the hydrological cycle, and water resources.

hydrometeors: Objects in the air made of water such as raindrops and snow crystals.

hydrosphere: The parts of the earth covered by water.

hypothermia: A dangerous fall in the temperature of an animal's body.

I

ice age: A period of thousands of years during which ice sheets cover many parts of the world.

iceberg: Ice that has broken off from an ice shelf or a mass of land-based ice and is floating in the water.

ice cap: A dome-shaped "iceberg" glacier spreading out in all directions, usually less than 50,000 square kilometers (approximately 19,000 square miles) in area.

ice core: A cylindrical length of ice removed from an ice sheet, ice cap, or glacier by a drill designed to do this.

ice sheet: A mass of glacier ice that covers the surrounding terrain and is greater than 50,000 square kilometers (approximately 19,000 square miles) in area.

ice shelf: Ice from a glacier, ice cap, or ice sheet that has pushed out over water and is floating but is still attached to its parent mass of ice.

ice storm: A storm with extensive freezing rain or drizzle that deposits ice on objects creating hazards to transportation and electrical power distribution.

infrared radiation (IR), infrared energy: **Electromagnetic radiation** with wavelengths from approximately 0.8 micrometers to 0.1 millimeters.

instrument flying: The control and navigation of an aircraft without reference to the horizon or the earth's surface.

insulator: A material that prevents the transfer of heat or electrical energy.

interglacial: An interval of tens or hundreds of thousands of years marked by mild climates between the glacial states of an ice age, such as the current period.

Intergovernmental Panel on Climate Change (IPCC): The World Meteorological Organization (WMO) body that provides objective information about climate change.

International Geophysical Year (IGY): The coordinated, international program of intense Earth sciences observations from July 1, 1957, to December 31, 1958.

Intertropical Convergence Zone (ITCZ): A narrow, discontinuous belt of **convective** clouds and thunderstorms paralleling the equator and marking the convergence of the trade winds of the Northern and Southern hemispheres.

Iñupiat: The native people of Alaska's Arctic Ocean coast and Bering Straits regions.

inversion: In meteorology, a departure from the usual decrease or increase with altitude of the value of an atmospheric property—usually temperature—and also the layer through which this departure occurs.

ionization: The gain or loss of electrons by atoms and molecules to form ions, which have a positive or negative electrical charge.

iridescence: In reference to clouds, brilliant spots, bands, or borders of colors (usually red and green) caused by the diffraction of light around water droplets.

isobar: A line showing equal atmospheric pressure on a weather map.

isotopes: Atoms of an element with the same number of protons in the nucleus but different numbers of neutrons.

J

jet stream: A relatively narrow river of very strong horizontal winds (usually 57 mph or faster) embedded in the general flow of planetary-scale winds aloft. They are generally located above fronts.

K

katabatic wind: A wind caused by cold, dense air flowing downslope. Can be a **bora wind**. Often used to distinguish a wind that does not result from the flow of wind over hills or mountains.

knot: As a measure of speed, a speed of one nautical mile an hour. A nautical mile is equal to roughly 1.15 statute miles.

L

lake effect snow: Snow formed downwind of large lakes or sometimes the ocean when cold air is warmed and moistened when it flows across relatively warm water.

La Niña: The phase of El Niño-Southern Oscillation oceanic and atmospheric pattern characterized by relatively cool water in the central Pacific.

latent heat: The heat released or absorbed as a substance changes among its solid, liquid, and gas phases.

lead: A long, linear area of open water in sea ice that may range from a few feet to more than a mile in width and tens of miles long.

lift: The force created by air flowing over an object, such as an airplane's wings.

lithometeors: Anything in the air made of mostly solid, dry particles such as dust and volcanic ash.

Little Ice Age: An extended time of relatively cold conditions in many regions of the globe from the middle of the thirteenth century to the middle of the nineteenth century with the coldest periods in the sixteenth and seventeenth centuries.

longitudinal wave: A wave in which the particles of the medium it's traveling through move back and forth in the direction of movement.

longwave radiation: In meteorology, the same as **infrared radiation**.

long waves: In meteorology, the three to five waves in global wind patterns that circle a hemisphere in the middle-latitude westerly winds.

low-level jet stream: A band of strong winds at an atmospheric level well below high-altitude jet streams.

M

meridional flow: In meteorology, a wind-flow pattern with a pronounced north-south component that roughly parallels meridians of longitude, as opposed to **zonal flow**.

meridional overturning circulation (MOC): Ocean currents driven by density differences and wind comprising a global system. Many oceanographers prefer this term to "thermohaline circulation," which has been commonly used.

mesocyclone: A vertical cylinder of rotating air, approximately 1 to 6 miles in diameter in a **supercell** thunderstorm.

mesoscale: In meteorology, events from a few miles to a few hundred miles across.

mesoscale convective complex (MCC): A persistent, nearly circular, organized cluster of many interacting thunderstorms covering an area of at least 100,000 square kilometers (36,627 square miles).

mesoscale convective system: An organized group of thunderstorms that is larger than individual thunderstorms, but smaller than **synoptic-scale** weather systems, and can last for hours.

mesoscale vorticity center: The circulation above the earth's surface associated with a mesoscale convective complex.

meteor: Phenomena that accompany a meteoroid as it passes through the atmosphere, including a flash and streak of light and an ionized trail.

meteoroid: A natural object from space that enters the atmosphere.

meteorologist: A scientist who specializes in atmospheric phenomena.

meteorology: The study of atmospheric phenomena.

microburst: An intense **downburst** from a shower or thunderstorm that affects an area (usually an oval, no longer than 2.5 miles) with strong and gusty winds.

microscale: In meteorology, weather events approximately a mile or less across.

mini or miniature supercell: A **supercell** thunderstorm with an **echo top** less than 10 kilometers (six miles) above the ground.

mistral (wind): A strong, cold, dry **katabatic wind** that flows down into the Rhone River Valley of France and on to the French and Italian Riviera on the Mediterranean coast.

mixing layer: The layer of the atmosphere below an inversion in which convection can uniformly disperse **air pollutants**.

mixing ratio: The ratio of the mass of water vapor to the mass of dry air, usually measured in grams of water vapor per kilogram of dry air.

model: A representation in any form of an object, process, phenomenon, or system designed to understand its behavior or to make predictions. Types include physical, conceptual, mathematical, and computer models.

Model Output Statistics: Statistical relations, derived over a long period of time, between locally observed weather elements and the output of a **numerical weather prediction model**.

moist adiabatic lapse rate: The rate at which saturated rising air cools as water vapor condenses, 2 to 3.5 degrees F per 1,000 feet.

N

nautical twilight: The interval of incomplete darkness before or after **civil twilight** when the sun's disk is between six and twelve degrees below the horizon. Barely enough natural light is available to distinguish the outlines of objects at ground level.

negative feedback: A process in which a change in a variable suppresses the process through interactions within the system. Opposite of **positive feedback**.

normal: In reference to climate, the average value of a weather element measured over thirty consecutive years.

North Atlantic Oscillation (NAO): An oscillation between relative atmospheric pressures of the Icelandic low and the Azores high linked to winds, storm tracks, precipitation, and temperatures across the North Atlantic Ocean.

Northern Annular Mode: See **Arctic Oscillation**.

numerical weather prediction model: A computer program designed to forecast the behavior of the atmosphere by using weather observations as input for mathematical models of the atmosphere, including its interactions with the earth's surface, to predict the future state of the atmosphere.

O

orbital wave: A wave in which the particles of the medium it's traveling through move in circular paths. Orbital waves transmit energy only along fluid boundaries such as between a liquid and gas or between areas of different densities in a liquid or in a gas.

ozone: An almost colorless, gaseous form of oxygen made of three oxygen atoms.

ozone layer: The general stratum of the upper atmosphere between roughly ten and forty miles above the surface with an appreciable ozone concentration and in which ozone plays an important part in the balance between incoming and outgoing radiation.

P

Pacific-North American Pattern: Oscillations in the strength and position of the jet stream and storminess over the eastern North Pacific and North America, which affect temperatures and precipitation across large parts of North America.

paleoclimatologist: A scientist who conducts research into ancient climates.

parameterization: In meteorology, a method of including the effects of processes, such as individual thunderstorms, that are too small to be represented on the grid of a **numerical weather prediction model**.

parts per million (ppm): A method of describing the proportions of very dilute concentrations of substances, such as in the air or water. For the air, such proportions are usually expressed in terms of volume and are abbreviated as ppmv.

passive remote sensing: The detection of energy, such as visible light, infrared energy, and microwaves, emitted or reflected by objects such as clouds, water vapor, and ice.

permafrost: Ground that remains at or below 32 degreesF for at least two years.

photochemical smog: A visibility-restricting, noxious mixture of aerosols and gases produced by reactions involving sunlight, oxides of nitrogen, and **volatile organic compounds**.

photometeors: Any luminous optical phenomena in the atmosphere, such as rainbows.

photon: In theory, the smallest quantity, or quantum, of electromagnetic energy.

phytoplankton: Microscopic unicelluar algae and bacteria that float or drift in the ocean and use solar energy to transform nutrients into the complex organic material that animals, including **zooplankton**, eat.

pineapple express: An informal term often used on the U .S. West Coast to describe an **atmospheric river**, often from the region around the Hawaiian Islands.

plate tectonics: The **theory** that the outer 60 miles of the solid earth is divided into a number of plates that move relative to one another over geological time.

polar easterlies: The irregular and diffuse surface east-to-west winds in the polar regions.

polar highs: Areas of high atmospheric pressure that form over the Arctic and Antarctic.

polar orbit: A satellite path that crosses both polar regions once during each orbit.

polynya: An irregularly shaped area of persistent open water in sea ice.

positive feedback: A process in which a change in a variable reinforces the original process. Opposite of **negative feedback**.

postdoctoral: Someone who has recently earned his or her PhD and is working in a temporary research post.

precipitable water: The total amount of precipitation if all of the water vapor in the air above a location condensed and fell.

precipitation: Water or ice that falls from a cloud to the ground.

pressure gradient force: The force that accelerates air from areas of high air pressure toward lower pressure. Most commonly used to refer to a horizontal force accelerating air to cause winds.

proxy climate sources: Sources of climate and weather data other than recorded weather observations.

pyrocumulus cloud: A towering cumulus cloud created by the rising hot air and smoke of a fire.

R

radiative forcing: A measure of the influence of a factor in altering the balance of incoming and outgoing energy in the earth-atmosphere system.

radiosonde: A small, balloon-instrument package that measures vertical profiles of the atmosphere's temperature, pressure, and humidity and radios the data to a weather station.

rain bands: Also called **spiral bands**; long, narrow lines of thunderstorms in a tropical cyclone that spiral in toward the center.

rear inflow jet: A mid-level flow of air into the rear of a **mesoscale convective system**.

refraction: The change in the direction of energy propagation as the energy moves into a medium with a different density such as the refraction of light that causes rainbows.

relative humidity: The ratio of the air's actual vapor pressure to its saturated vapor pressure expressed as a percentage.

resistor (electrical): A substance that resists the flow of an electrical current without blocking it. A circuit component designed to produce a certain amount of resistance.

return stroke (lightning): The intensely luminous discharge that propagates upward from the earth's surface to the cloud in a cloud-to-ground lightning flash.

ridge: In meteorology, an elongated area of high atmospheric pressure.

rime: A white or milky and opaque granular deposit of ice formed by the rapid freezing of supercooled water drops as they impinge upon an object.

rogue wave: An abnormally large wave compared with the other waves at the same time and place, possibly large enough to capsize or damage a ship.

Rossby waves: In meteorology, **long waves** in the middle-latitude westerly winds.

S

Santa Ana winds: A hot, dry **foehn wind** blowing down into the Los Angeles Basin. Often used to refer to similar winds elsewhere in Southern California.

saturated mixing ratio: The **mixing ratio**, usually measured in grams of water vapor per kilogram of dry air, when the air is saturated; that is, it contains the maximum amount of water vapor for its temperature.

saturation: In meteorology, generally the condition in which the partial pressure of water vapor in the air is equal to its maximum. That is, the relative humidity is 100 percent.

sea ice: Ice that forms when sea water freezes.

semipermanent highs and lows: Large areas of high or low atmospheric pressure that change in strength and extent with the seasons but remain generally in the same locations.

sensible heat: Heat absorbed or released by a substance when it changes temperature but does not change phase. See **latent heat**.

service ceiling: The highest altitude an airplane can reach while continuing to climb at a rate of at least 100 feet per minute, which is the aircraft's maximum usable altitude.

severe thunderstorm warning: An NWS warning that thunderstorms with winds of 58 mph or higher, or hail stones at least three-quarters of an inch in diameter, threaten the warned area.

shear zone: In oceanography, the zone along the edges of an ocean current where the current's movement creates swirls and eddies that concentrate **phytoplankton** and **zooplankton**.

shorefast ice: Sea ice that is frozen to the shore.

short wave: In meteorology, a short-wavelength ripple superimposed on a **long wave** in the upper level flow of air.

shower: Rain or snow that falls from **convective** clouds. Precipitation can be light or heavy, but it will quickly start and stop and can rapidly change intensity.

Southern Annular Mode: See **Antarctic Oscillation**.

Southern Oscillation: Opposing swings of surface atmospheric pressure between the eastern and western tropical Pacific Ocean, associated with **El Niño** and **La Niña** events.

spaghetti plot: A graphical method for displaying data from several model runs in an **ensemble forecast**.

spiral bands: See **Rain bands**.

squall line: Any non-frontal line or narrow band of thunderstorms.

stable (atmosphere): A layer of air characterized by a temperature gradient that causes air parcels to resist upward or downward forces. Contrast with **unstable (atmosphere)**.

statistical forecast: A forecast based on a systematic statistical examination of data representing past, observed behavior of the system to be forecast, including observations of useful predictors outside the system. This is in contrast to using a **numerical prediction weather model**.

steam fog: Any ground-level cloud produced when cold air comes in contact with a relatively warm water surface.

stepped-frequency microwave radiometer (SFMR): A **passive remote sensing** device used by aircraft to measure hurricane wind speeds and rainfall.

stepped leader: An ionized channel through the air from an area of electrical charge in a thunderstorm toward an area of positive charge that propagates by successive jumps. The first step in a lightning discharge.

stratosphere: The layer of the atmosphere above the **troposphere** and below the mesosphere, approximately 6 to 50 miles above sea level, where temperatures are generally constant with altitude in the lower levels and increase with altitude at higher levels.

stroke (lightning): Any one of a series of repeated electrical discharges composing a single lightning flash.

sublimation: In meteorology, the transformation of ice directly to water vapor. The opposite of **deposition**.

subpolar lows: Areas of **semipermanent low** atmospheric pressure located roughly between 50 degrees and 70 degrees latitude in both hemispheres.

subsidence inversion: The **inversion** that forms under a layer of air warmed as it descends.

subtropical highs: The **semipermanent** areas of high atmospheric pressure centered roughly on the 30 degrees north and south latitudes.

supercell: A long-lasting thunderstorm with a single, steady, intense **updraft** (usually rotating) and **downdraft**. Often produces severe weather including hail and **tornadoes**.

supercooled water: Water that is still liquid, not ice, at a temperature below 32 degrees F (0 degrees C).

supersaturated air: Air with a **relative humidity** above 100 percent.

sustained wind speed: In the United States, the wind speed obtained by averaging the wind speeds observed over one minute.

synoptic maps: Weather maps showing weather on the synoptic scale.

synoptic-scale weather: In meteorology, weather events from a few hundred to a thousand or more miles across.

T

teleconnection: A linkage between weather changes occurring in widely separated regions of the globe.

terminal aerodrome forecast (TAF): An aviation weather forecast for an airport and its vicinity.

terminal velocity: The maximum falling speed of any object when the forces of drag (air resistance) and gravity balance.

temperate climate: Average weather conditions in latitudes between the tropics and the polar regions.

temperate zone: The region of the earth between the tropics and polar regions.

theory: In science, unlike in common usage, a theory does not refer to speculation or opinion but to an overall explanation that links a large collection of observations or events. A scientific theory is based on hypotheses that have been verified many times by different

paper's daily weather page to reporting on ozone hole science from places as far away as the South Pole.

I also have to thank many people with three organizations for offering me the chance to watch scientists and forecasters at work over the years. Such a perspective is absolutely necessary for anyone who wants to write about science.

NOAA took me on flights into hurricanes aboard its WP-3 airplanes. NOAA's National Weather Service staff always made me welcome at forecast offices and forecasting centers, including the National Hurricane Center. (Being there when the eyewall of Hurricane Andrew battered the building was a hurricane education worth many books.) Scientists from NOAA and the National Center for Atmospheric Research welcomed me on tornado chases (I never caught one) and taught me much about severe weather. The National Science Foundation made dreams come true by inviting me to report on scientific work in Greenland, Antarctica, and aboard the U.S. Coast Guard research icebreaker, *Healy*, during an Arctic Ocean cruise. My few days aboard the *Healy* introduced me directly to the work of oceanographers. The time on the ship and my research beforehand amounted to a seminar on the role of the biosphere in the climate system. All of these experiences have informed this book.

Finally, my explorations of the worlds of weather, climate, polar, and other sciences would have been an empty journey without the loving companionship and encouragement of my wife, Darlene.

CREDITS

Photography:
The organizations or individual photographers listed below own all rights to the photographs used in this book.

About This Book: PP ii–iii, Mike Hollingshead.

Chapter 1: PP xii–1, FEMA, Jocelyn Augustino. P 14, Ken Tape. P 18, NOAA.

Chapter 2: PP 20–21, University Corporation for Atmospheric Research (UCAR), Lee Klinger. P 23, both, National Science Foundation (NSF), Peter West. P 26, UCAR, Carlye Calvin. P 41, Matt Mendelsohn. P 44, NSF, Robin Solfisburg.

Chapter 3: PP 46–47, FEMA, Greg Henshall. P 59, Bill Serne. P 64, Jessica Rinaldi. P 68, Robert Benson.

Chapter 4: PP 70–71, FEMA, Dave Gatley. P 73, NOAA Aircraft Operations Center. P 78, Carlye Calvin. P 79, UCAR. P 88, NSF, Bill Meurer. P 92, Matt Mendelsohn.

Chapter 5: PP 96–97, NOAA. P 100, Robert Benson. P 101, UCAR, Carlye Calvin. P 102, Darlene Shields. P 118, Matt Mendelsohn. P 122, Jack Williams.

Chapter 6: PP 124–25: NSF, Dwight Bohnet. P 126, *The Antarctic Sun*, Emily Stone. P 127, NSF, Steven Profaizer. P 128 top, *The Antarctic Sun*, Emily Stone. P 128 bottom, Jack Williams. P 130, Courtesy of Konrad Steffen. P 136, U.S. Air Force Reserve, T.Sgt. Ryan Labadens. P 148, American Meteorological Society, Curtis Compton.

Chapter 7: PP 150–51, Kevin Ambrose. P 158, Nate Billings. P 166, Anne Ryan. P 170, Matt Mendelsohn. P 171, UCAR.

Chapter 8: PP 176–177, Kevin Ambrose. P 180, Courtesy of The Parsons Company. P 185, Bruce Burkman. P 190, Anne Ryan. P 197, Bruce Burkman. P 200, UCAR, Carlye Calvin.

Chapter 9: PP 204–05, National Park Service, Jim Peaco. P 208, Troy Maben. P 220, David Sanders. P 227, NOAA National Environmental Satellite, Data, and Information Service (NESDIS).

Chapter 10: PP 228–29, NOAA NESDIS. P 232, Jack Williams. P 233, NOAA Aircraft Operations Center, Mike Silah. P 234, Mikki Harris. P 244, U.S. Air Force Reserve, Maj. Chad Gibson. P 247, NOAA Aircraft Operations Center, Dewey Floyd. P 255, Andrew Itkoff.

Chapter 11: PP 258–59, UCAR, Carlye Calvin. P 260, Jack Williams. P 262 both, Jack Williams. P 263, Matt Mendelsohn.

Chapter 12: PP 276–77, U.S. Antarctic Program, Zenobia Evans. P 284, Carlye Calvin. P 302, Matt Mendelsohn.

The global tropical cyclones map on Pages 256–257 was created for NOAA's International Best Track Archive for Climate Stewardship by Kenneth Knapp, Michael Kruk, David Levinson, NOAA employees; Howard Diamond, NOAA contractor; and James Kossin, University of Wisconsin.

GRAPHICS

The graphics in this book were drawn using Macromedia Freehand MX and Adobe Photoshop by artists Jeff Dionise and Rod Little, based on drawings and mock-ups by the author. Pages were composed using QuarkXPress 7.31 and 8.01.

ABOUT THE TYPE

The font used in this book is ITC Charter, designed by Matthew Carter in 1987. ITC Charter follows eighteenth-century Roman types but with three distinguishing design features: narrow proportions to allow for economical use of space; a generous x-height that improves readability; and sturdy, open letterforms legible in a variety of imaging environments.